Markus Krückemeier

# A Mobile Passive Radar System

with 69 figures

# A Mobile Passive Radar System

Von der Fakultät für Elektrotechnik, Informationstechnik, Physik
der Technischen Universität Carolo-Wilhelmina zu Braunschweig

zur Erlangung des Grades eines Doktors
der Ingenieurswissenschaften (Dr.-Ing.)

genehmigte Dissertation

von Markus Krückemeier
aus Minden

eingereicht am: 15.12.2021
mündliche Prüfung am: 13.05.2022

1. Referent: Prof. Dr.-Ing. Jörg Schöbel
2. Referent: Dr.-Ing. habil. Robert Geise
3. Referent: Prof. Dr.-Ing. Achim Enders

Druckjahr: 2022

**Dissertation an der Technischen Universität Braunschweig, Fakultät für Elektrotechnik, Informationstechnik, Physik**

**Bibliografische Information der Deutschen Nationalbibliothek**
Die Deutsche Nationalbibliothek verzeichnet diese Publikation in der Deutschen Nationalbibliografie; detaillierte bibliografische Daten sind im Internet über http://dnb.d-nb.de abrufbar.

1. Aufl. - Göttingen: Cuvillier, 2022
Zugl.: Braunschweig, Techn. Univ., Diss., 2022


Nonnenstieg 8, 37075 Göttingen
Telefon: 0551-54724-0
Telefax: 0551-54724-21
www.cuvillier.de



Gedruckt auf umweltfreundlichem, säurefreiem Papier aus nachhaltiger Forstwirtschaft.

ISBN: 978-3-7369-7674-0
eISBN: 978-3-7369-6674-1

*For Gabrielle and Emil. I am truly thankful to have you in my life.*

# Acknowledgements

This dissertation is the result of my work at the Institute of High Frequency Technology in the years 2014 to 2021. This work would not have been possible without the help and support of many people:
First of all, I would like to thank my supervisor Prof. Dr.-Ing. Jörg Schöbel, who made it possible for me to do this work part-time while working for the SF Microwave GmbH. I am grateful for the countless discussions, both about scientific topics and also about the bureaucratic tangles of project administration. I really enjoyed the time I spent in your research group.
Furthermore, I would like to acknowledge the other members of my examination board, Dr.-Ing. habil. Robert Geise and Prof. Dr.-Ing. Achim Enders, for taking the time to thoroughly review my dissertation and later discussing the finer details of passive radar signal processing with me.
Probably the best thing about the doctoral years was the opportunity to work with many great colleagues who have also become friends over the years. I am especially thankful to Dr.-Ing. Fabian Schwartau. We have worked together on a variety of projects over the years, and as a team we have always achieved more than two individuals could have. I am glad that we can continue this fruitful cooperation in our company in the future.
I would also like to thank my office colleague Sebastian Paul and Dr.-Ing Carsten Monka-Ewe for many hours of discussions and valuable advice in the field of numerical simulation. Of course, I am also grateful to all other IHF employees who have not been mentioned by name here; I have fond memories of many discussions, lunches, barbecues and celebrations. In addition to my colleagues at the institute, I would also like to thank Jan Fahlbusch, who provided guidance and advice on RF electronics design and layout during my time working at SF Microwave. In addition, many thanks to my good friend Carsten Bohnens, for many consultations, discussions and also the support in troubleshooting my software. Also, a big thank you to Dr.-Ing. Sebastian Brückner, who encouraged me to start the doctoral studies, without his incentive this book would not exist.
Furthermore, I would like to thank the many students I was privileged to supervise during my doctorate. There are two students in particular that I would like to point out, as their results have also been incorporated into this work. These were on the one hand Inna Frede, who dealt with target tracking and Kalman filters in her diploma thesis. On the other hand Marvin Wenzel, who delivered an outstanding work in the field of electronic beam steering with Rotman lenses showing an impressive attitude and methodology.
In addition, I want to thank the EU, which has financially supported my work on the ALFA project as part of the Horizon2020 program.
My deepest thanks go to my parents Rolf and Ingrid and my sister Britta for their constant support during my studies and the subsequent doctoral studies, without you I would not have made it this far.
Finally, I would like to thank my incredible fiancée Gabrielle and our wonderful son Emil. I look forward to tackling life's next challenges with you as my family.

*Markus Krückemeier*

# Contents

# Abbreviations

**ACAS** airborne collision avoidance system
**ADC** analog-to-digital converter
**ADS-B** automatic dependent surveillance - broadcast
**AGC** automatic gain control
**ALFA** advanced low flying aircrafts detection and tracking
**AMBOS** Abwehr von unbemannten Flugobjekten für BOS
**AoA** angle of arrival
**ATC** air traffic control
**AWGN** additive white Gaussian noise
**CA-CFAR** cell-averaging constant false-alarm rate
**CFAR** constant false-alarm rate
**CMA** constant modulus algorithm
**CNIT** national inter-university consortium for telecommunications
**CUDA** compute unified device architecture
**CUT** cell under test
**DFT** discrete fourier transform
**ECEF** earth-centered, earth-fixed
**EIRP** equivalent isotropic radiated power
**EKF** extended Kalman filter
**ENR** excess noise ratio
**ERP** Effective Radiated Power
**ETSI** European Telecommunications Standards Institute
**FFT** fast Fourier transformation
**FMCW** frequency-modulated continuous wave
**GNI** Gauß-Newton iteration
**GNSS** global navigation satellite system
**GUI** graphical user interface
**HPBW** half power beamwidth
**IFFT** inverse fast Fourier transformation
**IFR** instrument flight rules
**ISM** industrial, scientific and medical
**KF** Kalman filter
**LCMV** linearly constrained minimum variance
**LMS** least mean square
**LNA** low noise amplifier
**LO** local oscillator
**LOS** line of sight
**NLMS** normalized least mean square
**OS-CFAR** ordered-statistic constant false-alarm rate
**PA** planar approximation
**PCB** printed circuit board
**PLL** phase-locked loop
**radar** radio detection and ranging

| | |
|---|---|
| **RCS** | radar cross section |
| **RDS** | radio data system |
| **RF** | radio frequency |
| **rms** | root mean square |
| **SCA** | Subsidiary Communications Authorization |
| **SDR** | software-defined radio |
| **SMARP** | software-defined multiband array passive radar |
| **SNIR** | signal-to-noise and interference ratio |
| **SNR** | signal-to-noise ratio |
| **SUT** | scaled unscented transform |
| **TCXO** | temperature compensated crystal oscillator |
| **UAV** | unmanned aerial vehicle |
| **UCA** | uniform circular array |
| **UKF** | unscented Kalman filter |
| **ULA** | uniform linear array |
| **VCO** | voltage-controlled oscillator |
| **VFR** | visual flight rules |
| **VHF** | very high frequency |
| **W-LAN** | wireless local area network |

# 1. Introduction

Nowadays, the detection of distant objects by means of electromagnetic waves is usually described as radio detection and ranging (radar). The basic technique can be summarized relatively simply. One example for such a summary can be found in the radar handbook by Merill Skolnik [1]:

> *"Radar is an electromagnetic sensor for the detection and location of reflecting objects. Its operation can be summarized as follows:*
>
> - *The radar radiates electromagnetic energy from an antenna to propagate in space.*
> - *Some of the radiated energy is intercepted by a reflecting object, usually called a target, located at a distance from the radar.*
> - *The energy intercepted by the target is reradiated in many directions.*
> - *Some of the reradiated (echo) energy is returned to and received by the radar antenna.*
> - *After amplification by a receiver and with the aid of proper signal processing, a decision is made at the output of the receiver as to whether or not a target echo signal is present. At that time, the target location and possibly other information about the target is acquired."*

The technology goes back to the first fundamental experiments of Heinrich Hertz in 1886 [2]. He demonstrated the electromagnetic waves predicted 20 years earlier by James Clerk Maxwell and performed experiments on their reflection from metal plates - the first radar measurement. The first technical application of this effect followed with the "Telemobiloscope" of Christian Hülsmeyer, who demonstrated a device capable of detecting ships on the Rhine in Cologne in 1904 [3]. The first large-scale air surveillance radar, "Chain Home," was developed in the years before World War II by Robert Watson Watt and Arnold Wilkins. Their Davenport experiment in 1935 proved the technology could be used on a larger scale [4].

In addition to the classic areas in maritime and aviation surveillance, radar is nowadays used in a wide variety of applications: Weather radar systems are used to study the atmosphere, ground-based radar systems are used to monitor space debris, satellite-based radar systems are used for earth observation and to study other celestial bodies, ground penetrating radar systems are used in geology and archaeology, in the automotive sector radar sensors are used for driver assistance systems and many more examples of applications could be found. As diverse as the areas of application are, as diverse are the systems and challenges involved. The further development and improvement of radar is therefore still a topic of current research. The current reasearch topics start with the improvement of the underlying hardware, i.e. components such as signal sources, amplifiers, mixers, and antennas and also include questions about the backscattering properties of the potential targets. But, mainly they deal with the advancement of signal processing methods [5,6].

In this thesis, a passive radar system designed for ground-based surveillance of airspace is presented. The concept of a passive radar is to use an illuminator of opportunity instead of transmitting a signal on its own. This principle was already used in the early days of radar. In fact, the already mentioned Davenport experiment of Wilkins used a public radio transmitter as the illuminator and therby classifies as a passive radar experiment.

However, further development of this technology was initially postponed in favor of the much easier-to-use monostatic radar systems, which transmit and receive from the same location[1]. It was not until the 1980s that passive detection came back into focus. Since the late 1990s and early 2000s, the first commercially available systems have been the Silent Sentry systems from Lockheed-Martin. These system were followed by other systems, such as the Homeland Alerter from Thales, the AULOS system from LEONARDO, a system from Airbus Defence and Space, and several other commercial systems [8, 9]. In parallel to these systems, a number of systems have also been developed for research purposes. Some examples are the PaRaDe system of the Warsaw University of Technology [10], various systems of the Fraunhofer Gesellschaft, such as Cora [11], or the software-defined multiband array passive radar (SMARP) system of the Italian national inter-university consortium for telecommunications (CNIT) [12]. When the research for this thesis was started, the supply of reference books on this research area in particular was still comparatively thin, however, in the meantime textbooks on this subject have been published. The first book on this topic was "An Introduction to Passive Radar" by Hugh D. Griffith in 2017 [9]. This was followed by the book "Signal Processing for Passive Bistatic Radar" by Mateusz Malanowski in 2019 [13]. These books describe many of the fundamentals used in this work in detail and are recommended as further reading.

The question, on which this thesis is based, resulted from the work with different projects around the detection of small aircraft, both manned and unmanned. E.g., an active frequency-modulated continuous wave (FMCW) radar system was developed for the detection of small air vehicles, which is also briefly described in the Appendix A. The first experiments in the field of digital beamforming and subsequent signal processing were carried out with this system.
The passive radar system described in this thesis was developed within the project: advanced low flying aircrafts detection and tracking (ALFA) [2]. This project was concerned with the detection of low-flying small aircraft or unmanned aerial vehicles (UAVs), which are frequently used for drug smuggling at the European southern border. ALFA relied on the combination of an active radar system, an electro-optical sensor, and the passive system described here. The measurements of these systems were merged via a sensor data fusion component. The system was additionally used as a passive radio detector, primarily for the detection of UAV. Although this type of use is not the subject of this thesis, it has been investigated and a short overview can be found in the Appendix D.

[1]Such monostatic radar systems are currently used for air traffic control. With the Ramet MSSR M10SR [7] as an example, here are a few key figures on the performance of such systems currently in use: maximum Range 256 NM, update rate 4 - 7.5 sec, angular mean square error 0.1 deg, range mean square error 60 m, altitude information None, detection probability 98 %

[2]The ALFA project has received funding from the European Union's Horizon 2020 programme under grant agreement No. 700002.

## 1.1. Contributions

The contribution of this work to the state of the art can be summarized in the following items:

- **SDR based Passive Detection System**
  As part of the ALFA [14] project, a combined system for passive detection and passive radar was built. The system includes a modular antenna system, receiver hardware and signal processing hardware. The system is used for further investigations within this thesis and represents the base for additional research in this area. The developed system is presented in Chapter 3.
- **Illuminator Locator**
  In Chapter 4.3, different methods for determining the illuminator position in bistatic passive radar systems are compared. The planar approximation (PA) method is introduced as an approach to improve the accuracy of iterative solution techniques. The method is compared with previous approaches and is found to be superior in the scenario studied.
- **PA based Tracking Algorithm**
  In Chapter 4.4, a tracking method is proposed that solves the fusion of measurements for track initialization with the PA method. The method can be used to determine targeted hypotheses for new tracks, reducing the effort required to find the right combinations of targets from different illuminators.
- **14 GHz Collision Avoidance Radar**
  Appendix 2.3 is an overview of the prototype of a 14 GHz FMCW collision avoidance radar that was developed as part of this thesis. Furthermore, exemplary results from student work are shown, in which signal processing and electronic beam steering for this system are demonstrated.

## 1.2. Organization of the Thesis

Following this introductory chapter, this dissertation is organized as follows:

**Chapter 2** provides a selection of the important fundamentals of the radar systems used. The concept of radar is considered in general; the operation of FMCW radar, as well as the concept of passive radar are explained. The ambiguity function, which is relevant for both systems, is also introduced. In the field of antenna theory, beamforming and the modeling of antenna arrays are described. The explanations of the Kalman filter, as well as some nonlinear extensions of it conclude this chapter.

**Chapter 3** describes the coherent receiver system that was developed as part of this thesis. The chapter starts with an overview of the used hardware, as well as the synchronization of the receive channels. In further sections, the antennas and the measurement of the noise figure of the system are described.

**Chapter 4** describes the passive radar system developed in this thesis. The considerations for the selection of the illuminator system are described, as well as the complete signal processing. A separate section describes a method for determining transmitter positions using cooperative targets. Furthermore, in addition to the basic signal processing of the measured data, the fusion of this information by means of a tracking algorithm is described. This chapter concludes with the development of a model for the performance of the system.

**Chapter 5** presents measurements performed with the developed system. The results of the system are compared with the model, as well as with independent reference data. The chapter concludes with a discussion of the results.

**Chapter 6** summarizes the thesis and recommends future work.

**Appendix A** provides a brief overview of an active, FMCW radar system that was built early in this research. In the context of this work, it represents an experimentation platform on the basis of which initial experiments in signal processing were carried out.

**Appendix B** describes an alternative method for synchronizing the software-defined radio (SDR) channels in which only the common reference signal, which is distributed to all channels, is used and the local oscillator (LO) frequencies are generated with the internal phase-locked loops (PLLs). The limitations of this technique that led to its omission are also explained.

**Appendix C** is a compilation of measurement and simulation results for the characteristics of the amplifiers and antennas used.

**Appendix D** gives a brief outlook on the application of the system for the passive detection of radio signals. This technique is of high interest in the context of detection of smallest UAVs, but in the context of this thesis it is a proof of concept and functional verification of the developed hardware.

# 2. Fundamentals

This chapter covers a selection of the basics used in this work. They are intended to give the reader a brief overview on the topics and terminology used. If necessary, there are references to further literature given.

## 2.1. Radar Fundamentals

A radar system uses electromagnetic waves to detect distant objects by their reflection. The basic idea of a radar system can be explained as follows: An electromagnetic wave is radiated. After a certain propagation time $t$, the wave hits an object in a distance $R$ where it is reflected and/or scattered [1]. After another time $t$, the reflected wave can be detected at the receiver. From the propagation time $2t$ of the signal and the known propagation speed $c$, the distance of the object can be determined.

The range of a radar system depends on whether the backscattered signal can still be detected at the receiver. To determine this, first the magnitude of the received power is described. Of course, this largely depends on the transmit power $P_\mathrm{I}$ with which the radar illuminates the target. Assuming that the power is distributed isotropically, i.e., spherically around the transmitter, the radiation density at distance $R_1$ is the power of the transmitter distributed over the surface of a sphere $4\pi R^2$.

Considering that the transmitter does not radiate the power isotropically but directed with an antenna, the directivity of the antenna can be understood as an increase of the transmitted power by the factor $D$ in the direction of the radiated beam. With a real antenna, additional losses occur, which are described by the efficiency $\eta$. These quantities are summarized as the gain of the antenna $G_\mathrm{I} = \eta D$.

At the target, a certain part of the radiation density is reflected back in the direction of the receiver. Furthermore, directional effects and losses can also occur during this process. These effects are summarized as the effective aperture $\sigma$ of the target, which is also referred to as the radar cross section (RCS). It is given in units of area.
Due to the various effects summarized in the RCS, it depends on the angle at which the target is illuminated and can vary greatly for different directions of incidence[2]. The radiation density at the target and the RCS $\sigma$ can again be modeled as isotropically radiated power originating from the location of the target. On the way back to the receiver, the power is therefore again distributed on a spherical surface with the radius $R_2$. At the receiver the power density is then received with the effective antenna area $A_\mathrm{W}$. This can be determined from the wavelength $\lambda$ of the signal and the gain of the receiving antenna $G_\mathrm{R}$ [15, p.227] as

$$A_\mathrm{W} = \frac{\lambda^2}{4\pi} G_\mathrm{R}. \tag{2.1}$$

[1] Both terms are used synonymously. The exact physical process is not relevant for the purpose of this study.

[2] These effects result from the complex geometry of the targets and the materials used.

In summary, the power received at the radar is given by

$$P_{\mathrm{R}} = P_{\mathrm{I}} G_{\mathrm{I}} \cdot \frac{1}{4\pi R_1^2} \cdot \sigma \cdot \frac{1}{4\pi R_2^2} \cdot \frac{\lambda^2 G_{\mathrm{R}}}{4\pi}. \tag{2.2}$$

Assuming, that receiver and transmitter are in the same place[3] ($R = R_1 = R_2$ ), this expression can be reformulated into the so-called radar equation, which can be used to specify the maximum range of the radar as a function of the minimum signal power $P_{\mathrm{R,min}}$ that can be detected by the receiver, for a target defined by its RCS:

$$R_{\mathrm{max}} = \sqrt[4]{\frac{P_{\mathrm{I}} G_{\mathrm{I}} \sigma \lambda^2 G_{\mathrm{R}}}{(4\pi)^3 P_{\mathrm{R,min}}}}. \tag{2.3}$$

## 2.2. Topologies

An important aspect of the classification of radar systems is their topology. The most common are so-called monostatic radar systems. In a monostatic radar system is the transmitting and receiving units are located at the same site. In contrast, there are bistatic systems, where the transmitter and receiver are located at different sites, or multistatic systems, where multiple transmitters and/or receivers at different sites are used. An example of a bistatic topology is shown in Figure 2.1 for a passive radar system [1, 16]. The distance between illuminator **I** and observer **B** is defined as

$$R_{IB} = \|\vec{IB}\| \tag{2.4}$$

and is often referred to as the baseline. In addition, the distances between the illuminator **I** and the target **T**

$$R_T = \|\vec{IT}\| \tag{2.5}$$

as well as the distance between the receiver **B** and the target **T**

$$R_B = \|\vec{TB}\| \tag{2.6}$$

are also relevant.

## 2.3. FMCW Radar

Besides their topology, radar systems can also be distinguished by the type of illumination signal they use.

The signal shape is decisive for the information that can be obtained about a target. For example, in the early radar systems, only the distance to a target was determined. A simple pulsed signal was used for this purpose. With such a pulsed radar, the propagation time of a single pulse is measured for the round trip to the target. Such a system therefore provides information about the propagation time and thus about the distance to the target. For other radar systems, however, the information about the movement of an object is more relevant. As an example, an unmodulated continuous wave radar can resolve distances only in the range of a single wavelength by measuring the phase of the returned signal

[3] The scenario is therefore a monostatic one.

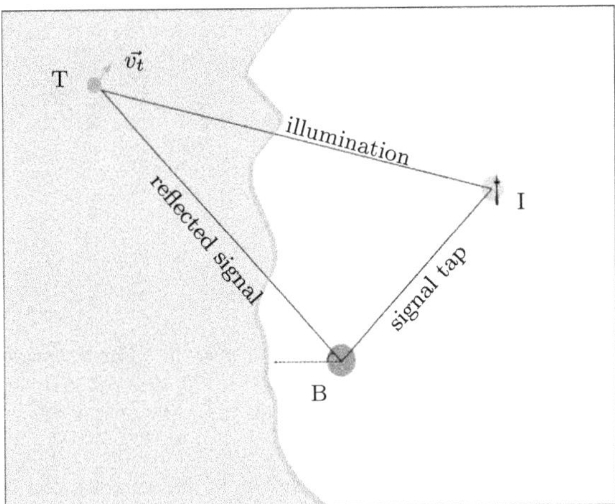


Figure 2.1.: Passive radar concept. A target T gets illuminated by an illuminator of opportunity at I. The observer B receives the direct signal from I and the reflected signal from T.

compared to the transmitted signal. This results in ambiguity of the received signal with a period of one wavelength. Therefore, these are usually unsuitable for distance measurement. However, from the phase shifts due to the Doppler effect, conclusions can be drawn about the radial velocity of the target. The Doppler shift results from the radial velocity $v_\mathrm{r}$ and the wavelength $\lambda$ of the signal [15, p. 69]:

$$f_{\mathfrak{D}} = \frac{2v_\mathrm{r}}{\lambda}. \tag{2.7}$$

One possibility to acquire both range and Doppler information of a target is the triangular **modulated** FMCW radar. The description of the FMCW radar is used here as a simple example of a (comparatively simple) modulated system and its evaluation for range and Doppler data. As the name implies, this is a radar in which the target is illuminated with a frequency modulated constant carrier. A good description, on which also the explanations shown here are based, can be found either in [15, p. 82 - 92] or in [17].

A simplified concept of such a radar is shown in Figure 2.2. The modulated radar signal is generated with a signal source and then split between the transmit antenna and a frequency mixer. At the mixer, the undelayed signal is used as the LO and mixed with the signal reflected at the target. The reflected signal is delayed by the signal propagation time. The Mixing of these signals results in the so-called beat frequency $f_\mathrm{b}$. The relationship between the transmitted and the received signals is shown in Figure 2.3 for a distant, moving target. The transmitted signal is shown in blue; the received signal in green. The received signal is delayed by the propagation time and shifted in frequency by the Doppler effect. These shifts result in two distinct beat frequencies for the up- and the down-sweep. The difference

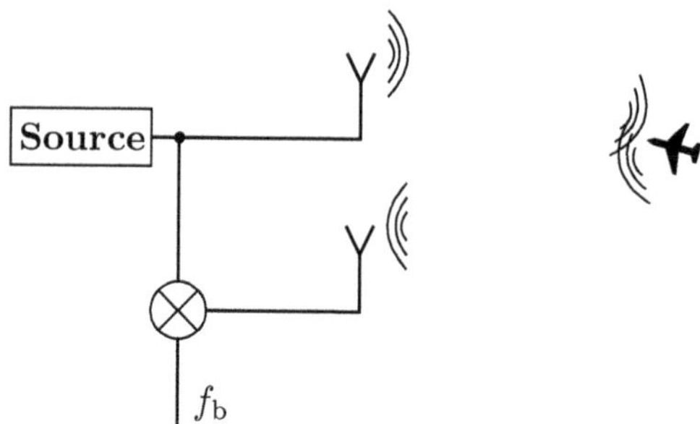


Figure 2.2.: Simplified diagam of an FMCW radar system. The generated radar signal is transmitted and reflected at the target. The delayed received signal is mixed with the signal from the source, resulting in the beat frequency $f_b$.

between the two is caused by the Doppler effect alone[4]. Therefore, the Doppler frequency of the target can be determined from the average difference between these two signals

$$f_{\mathfrak{D}} = \frac{f_{\mathrm{b,up}} - f_{\mathrm{b,dn}}}{2}. \tag{2.8}$$

The propagation time resulting from the target's distance also results in a beat frequency. If the target is stationary, this delay of the signal results in the same beat frequency for the up- and downsweep signal. However, the frequency shift from the Doppler effect is superimposed on this effect. The beat frequency is shifted by the same amount in each of the two sweeps, but in opposite directions in each case. Due to the equal but opposite shift, the frequency component resulting from the distance of the target $f_{\mathfrak{R}}$ can be determined as the average of the two signals:

$$f_{\mathfrak{R}} = \frac{f_{\mathrm{b,up}} + f_{\mathrm{b,dn}}}{2}. \tag{2.9}$$

Due to the time delay of the received signal, an ambiguous area is created between the ramps, in which the reflected signal of the last ramp is mixed with the signal of the new ramp. The size of this area depends on the distance of the target. Of course, this must be taken into account for signal processing, by not evaluating potentially ambiguous areas of the measurement. To identify these areas, the maximum range of the radar can be estimated with Equation (2.3). The propagation time of the signal for this range is then given by

$$\delta t_{\mathrm{ambigous}} = \frac{2R_{\mathrm{max}}}{c}. \tag{2.10}$$

The range resolution of an FMCW radar depends on the bandwidth $B$ of the sweep used and is given by

$$\Delta\mathfrak{R} = \frac{c}{2B}. \tag{2.11}$$

In addition, when processing the signal from such an FMCW radar, a process gain is achieved, since the **correlated** energy of the target is integrated over the entire usable sweep duration and bandwidth, while the **uncorrelated** energy of the noise remains evenly

[4]The considerations here apply when an ideal, non-complex mixer is used.

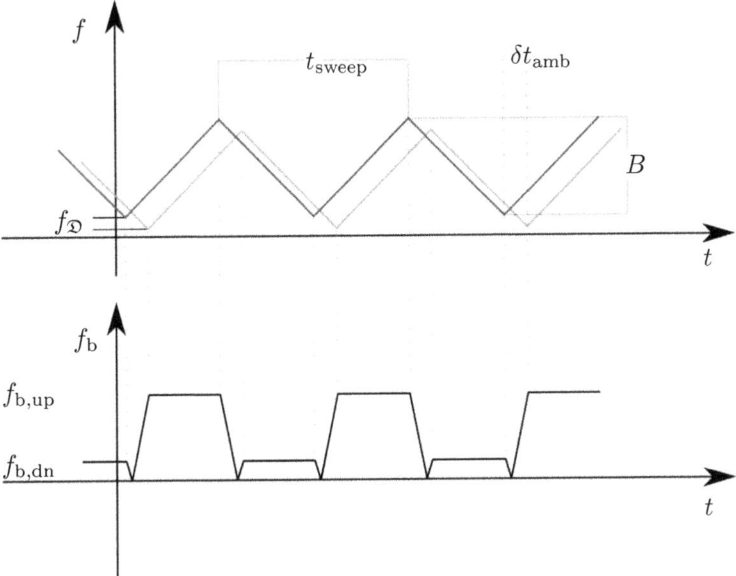


Figure 2.3.: Principle of a triangular modulated FMCW radar. The upper plot shows the frequency over time for the transmit signal (blue) and the receive signal (green) for a distant, moving target. The lower plot shows the beat frequency resulting from the complex mixing of these two signals.

distributed over the bandwidth. Thus, it can be calculated using the effectively used sweep bandwidth $B_{\text{eff}}$ and the effective sweep time. Since upsweep and downsweep are evaluated separately, they must also be considered separately. The effective sweep time for the individual sweeps is therefore calculated as

$$t_{\text{eff}} = \frac{t_{\text{sweep}}}{2} - \delta t_{\text{ambigous}}, \tag{2.12}$$

where $t_{\text{sweep}}$ is the absolute time for upsweep and downsweep (compare figure 2.3). The process gain for the individual sweeps is then calculated with

$$G_{\text{process}} = B_{\text{eff}} \cdot t_{\text{eff}}. \tag{2.13}$$

As already shown in Equation (2.3), the range depends on the minimum power the system can resolve $P_{\text{R,min}}$. This power can be calculated depending on two parameters. The first is thermal noise, which can be described by

$$P_{\text{N}} = k_{\text{B}} T_0 B_{\text{eff}}, \tag{2.14}$$

where $k_B \approx 1.38 \cdot 10^{-23} \frac{J}{K}$ is the Boltzmann constant and $T_0$ the equivalent noise temperature of the antenna. The second factor, which limits the detectable power, stems from additional noise contributions that arise in the system. These are described by the system's noise figure $F$.

With the process gain and the description of the noise, Equation (2.2) for the received power can be extended to get a relation between signal power and noise power. This eliminates the bandwidth from this equation, resulting in

$$\frac{P_R G_{\text{process}}}{P_N} = P_I G_I \cdot \frac{1}{4\pi R_1^2} \cdot \sigma \cdot \frac{1}{4\pi R_2^2} \cdot \frac{\lambda^2 G_R}{4\pi} \cdot \frac{t_{\text{eff}}}{k_B T_0} = \text{SNR}. \tag{2.15}$$

This ratio is called the signal-to-noise ratio (SNR). It is an important measure in assessing signal quality, the larger this ratio is, the farther the wanted signal's power is above the noise.
Rearranging this expression for the range $R$ yields a variant of the radar equation dependent on the system parameters

$$R = \sqrt[4]{\frac{P_I G_I \lambda^2 G_R t_{\text{eff}} \sigma}{\text{SNR}(4\pi)^3 k_B T_0 F}}, \tag{2.16}$$

with $R = R_1 = R_2$.

## 2.4. Passive Radar

A passive radar system uses an illuminator of opportuinty, which is not directly part of the system itself. Therefore, such systems usually have a bi-static topology; Figure 2.1 on page 7 shows an exemplary setup of such a system.
An independent transmitter **I** is used to illuminate the target **T**. The system's receiver **B**, located at another position, receives the reflected signal and the direct signal. By correlation with the transmitted signal (and its Doppler-shifted variations), the bi-static range $\mathfrak{R}$ and the bi-static Doppler shift $\mathfrak{D}$ is obtained. Unlike in mono-static topologies, $\mathfrak{R}$ does not directly express the distance to the target but the difference in path length between the direct line of sight (LOS) to the illuminator and the reflected signal. $\mathfrak{R}$ is calculated from the time difference of arrival between the reflected signal and the LOS signal tap. It is defined as [9]

$$\mathfrak{R} = \|\vec{BT}\| + \|\vec{TI}\| - \|\vec{BI}\|. \tag{2.17}$$

The bi-static Doppler frequency is calculated from the frequency shift resulting from the target's velocity $\vec{v}$ and the center frequency of the illuminating signal $f_I$. The Doppler frequency $\mathfrak{D}$ for this scenario consists of the dot product of $\vec{v}$ with the difference of the two radial components, one from the Illuminator **I** to the target **T** and one from the target to the observer **B**. It is defined by [18]

$$\mathfrak{D} = f_I \cdot \left( \frac{\vec{v}}{c} \cdot \left( \frac{\vec{TB}}{\|\vec{TB}\|} - \frac{\vec{IT}}{\|\vec{IT}\|} \right) \right). \tag{2.18}$$

This equation is valid for $\|\vec{v}\| \ll c$.

The most crucial factor in assessing a radar system's performance is the SNR achievable for a given target. This results from the ratio of the received power $P_R$ and the contributions

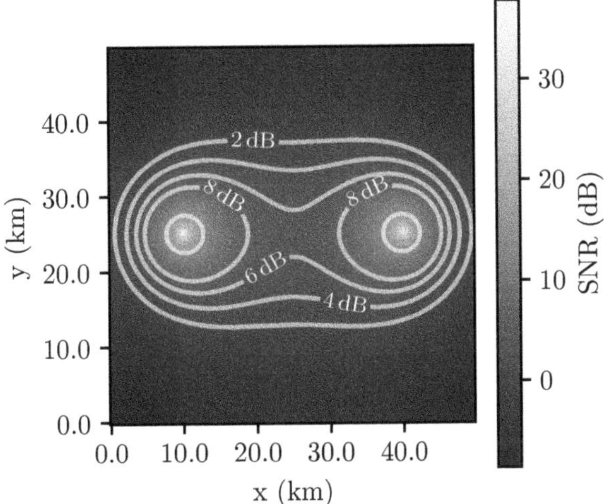


Figure 2.4.: Distribution of the SNR in a bistatic scenario.

of the noise[5] $P_{\mathrm{N}}$. The received power of the system is given by Equation (2.2), where $R_1 = R_{\mathrm{T}}$ represents the distance between the transmitter and the target and $R_2 = R_{\mathrm{B}}$ represents the distance from the target to the receiver.

The power of the noise $P_{\mathrm{N}}$ is described by the thermal noise according to Equation (2.14) and the additional noise from the system, described by its noise figure $F$. Put together, the SNR can be formulated as

$$\mathrm{SNR} = \frac{P_{\mathrm{R}}}{P_{\mathrm{N}}} = P_{\mathrm{I}} G_{\mathrm{I}} \cdot \frac{1}{4\pi R_T^2} \cdot \sigma \cdot \frac{1}{4\pi R_R^2} \cdot \frac{\lambda^2 G_{\mathrm{B}}}{4\pi} \cdot \frac{1}{k_B T_0 B F}, \tag{2.19}$$

which is valid for $T_0 = 290\,\mathrm{K}$.

Figure 2.4 shows the distribution of the SNR plotted over the position of a target for such a bistatic scenario. The transmitter is assumed to be at ($x = 10\,\mathrm{km}$ $y = 25\,\mathrm{km}$), the receiver at ($x = 40\,\mathrm{km}$ $y = 25\,\mathrm{km}$). The transmit power is assumed to be $P_{\mathrm{I}} = 1\,\mathrm{kW}$ and a gain of $G_{\mathrm{I}} = G_{\mathrm{B}} = 3\,\mathrm{dB}$ for the transmit and receive antennas. The RCS is assumed to be constant for this example with $\sigma = 10\,\mathrm{m}^2$; the signal is modeled with a wavelength of $\lambda = 3\,\mathrm{m}$ and a bandwidth of $B = 100\,\mathrm{kHz}$. The noise is calculated with $T_0 = 290\,\mathrm{K}$ and a system noise figure of $F = 3\,\mathrm{dB}$. The representation shows a high SNR in the direct vicinity of the receiving antenna as well as in the vicinity of the transmitting antenna. In the area between the transmitter and receiver, the minimum SNR is exactly in the center. Outside this area, the contours of constant SNR can be described with Cassini ovals [19].

[5] In a real system, interference e.g. due to a direct coupling of the reference signal also has an effect, those effects will be considered in more detail later in Chapter 4.5.

The distribution of the theoretically achievable SNR shows the greater complexity compared to monostatic systems, where a circular distribution of the SNR would result. This complexity is also reflected in the description of the achievable resolution of such a system. For the resolution of the bistatic range $\mathfrak{R}$ only the bandwidth of the used signal is relevant at first. This results in

$$\Delta\mathfrak{R}_{\text{min}} = \frac{c}{B}, \tag{2.20}$$

where $c$ describes the propagation speed of the signal and $B$ its bandwidth.
However, for the actual distance resolution between two targets $\Delta\hat{\mathfrak{R}}_{\text{min}}$, the bistatic geometry of the scenario must be considered [9, p. 36ff.]. With the geometry of the scenario that is depicted in Figure 2.5 the following equation can be derived:

$$\Delta\hat{\mathfrak{R}}_{\text{min}} = \frac{\Delta\mathfrak{R}_{\text{min}}}{\left[2\left(\cos\frac{\beta}{2}\right)\right]\cos\varphi} \tag{2.21}$$

Here $\beta$ is the bistatic angle defined between the lines from the target to the transmitter and receiver. The lines bisecting $\beta$ are called the bistatic bisector. The angle $\varphi$ is defined between the bistatic bisector and the direct line to the next target.

A similar approach must be taken to determine the velocity resolution. First, the resolution with which the Doppler frequency $\mathfrak{D}$ can be measured must be considered. This resolution depends only on the measuring time $T$:

$$\Delta\mathfrak{D}_{\text{min}} = \frac{1}{T}. \tag{2.22}$$

According to Equation (2.18), the Doppler frequency depends only on the velocity components projected onto the bistatic bivector. Therefore, only this component of the speed can be considered in the system. The difference of these velocity components $\Delta\mathfrak{V}$ of two targets located at the same position can be determined as follows:

$$\Delta\mathfrak{V} = (V_1\cos\delta_1 - V_2\cos\delta_2) \tag{2.23}$$

where $\delta$ expresses the respective angle between the bistatic bivector and the velocity vector. The resolution with which this bistatic velocity component can be measured depends also on the frequency of the illuminator signal $f_{\text{I}}$, respectively its wavelength $\lambda_{\text{I}}$:

$$\Delta\mathfrak{V}_{\text{min}} = \frac{c\cdot\Delta\mathfrak{D}_{\text{min}}}{f_{\text{I}}} = \frac{\lambda_{\text{I}}}{T} \tag{2.24}$$

If bistatic geomery is also included here, the result is the formula that can also be found in [9, p.40]:

$$\Delta\hat{\mathfrak{V}}_{\text{min}} = \frac{\Delta\mathfrak{V}_{\text{min}}}{\left[2\cos\left(\frac{\beta}{2}\right)\right]}. \tag{2.25}$$

Further details can be found in the books by Griffith [9] and Malanowski [13].

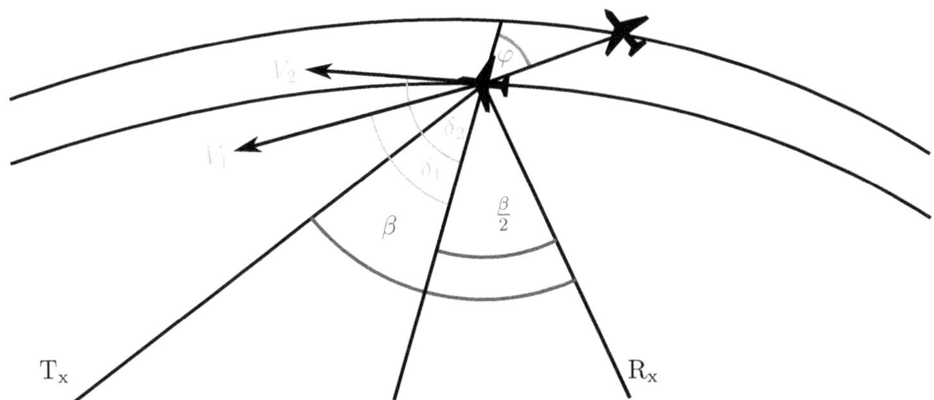


Figure 2.5.: 2D section of the geometry for bistatic range and Doppler resolution. The bistatic angle is shown in blue. In red the angle between two close targets is shown, which is included in the range resolution. In green the angles of the velocity components of two targets at the same location are shown for the calculation of the Doppler resolution.

## 2.5. The Ambiguity Function

The resolution properties of a (passive) radar system are determined by the waveform it uses to illuminate the targets. From equations (2.21) and (2.25), it can be seen that the resolution of the bistatic range is determined by the effective bandwidth of the signal while the resolution of the bistatic velocity is determined by the integration time. Since these two properties can be chosen (quite) independently, it can be assumed that the larger the bandwidth $B$ and the longer the integration time $T$, the better the performance of the radar (see [15, p. 409]). However, in addition to the pure resolution, the uniqueness of the result must also be guaranteed. As an illustrative example of this problem, one can imagine a time-continuous sinusoidal signal being used for the determination of a distance. For this purpose, the transmitted signal is compared with the received, time delayed signal. Within a wavelength, the difference in time of flight and thus the distance can be determined very precisely from the phase deviation between the reflected signal and the received signal. But distances, which correspond to a multiple of a complete wavelength, cannot be distinguished from each other. Periodicity in the time signal thus leads to ambiguities in the measurement of the range. Similarly, one can imagine that periodicity of the transmitted signal in the frequency domain leads to ambiguities in the Doppler measurement. These ambiguities of the transmitted signal can be investigated with the so-called ambiguity function. These thoughts go back to the first considerations of J. Ville in 1948 [20] and in the context of radar signal processing were formulated for the first time by P.M. Woodward [21] in 1953. Woodward defines a "correlation function" which Spafford specifies as a "time-frequency cross-correlation function" [22]:

$$\mathcal{X}(\tau,\vartheta) = \int u(t) \cdot u^*(t+\tau) \cdot \exp\left(-2\pi j \vartheta t\right) dt, \tag{2.26}$$

where $u(t)$ represents the examined radar signal, $\tau$ describes the delay of this signal and $\vartheta$ the frequency shift. In this definition it is assumed that the signal energy is normalized to unity and implicitly the same impedance is assumed for all signals

$$\int |u|^2 \, dt = 1. \tag{2.27}$$

The ambiguity function described (among others) by Skolnik [15] is then defined as the magnitude square of Equation (2.26)

$$\psi(\tau, \vartheta) = |\mathcal{X}(\tau, \vartheta)|^2. \tag{2.28}$$

In this definition, the ambiguity function is proportional to the signal power, while the correlation function is proportional to an amplitude value.

On first sight, this is not intuitive, since the result of (2.26) already obviously corresponds to energy and thus should already be proportional to power. In [1,15] Skolnik describes that in the general case, the maximum value of the ambiguity function is $(2E)^2$ with $E$ being the energy of the echo signal. Of course, Skolnik's description is absolutely correct, but it easily leads to the misinterpretation that the results from Equation (2.26) can already be treated as proportional to a power value.
For better understanding, the function should be interpreted as a matched filter, which is closely related to the ambiguity function, as described by Spafford in [22]. An even more generalized way is applying the condition in (2.27), which yields a normalized correlation (cf. [23, p. 204ff. and p. 238ff.]). All three approaches differ in principle only in whether none, one or both terms of the signal $u(t)$ are normalized in accordance with the requirement from (2.27). The three cases are proportional to each other, and their result is a measure of the similarity of the input signal to the reference signal.

Since this work deals mainly with sampled signals, the ambiguity function is defined as follows:

$$\psi(m, k) = \left| \sum_{n=0}^{N-1} s_{\mathrm{Ref}}(n) \cdot s^*_{\mathrm{Ref}}(n + m) \cdot \exp\left(\frac{-2\pi jkn}{N}\right) \right|^2. \tag{2.29}$$

Where N is the number of signal samples, $m$ is an integer number of the range bin delay and $k$ is an integer number of the Doppler bin shift. The size of $N$ is directly dependent on the integration time $T$ and the sampling frequency $f_\mathrm{s}$.

The defined ambiguity function has two applications: First, it is used to examine radar signals for inherent ambiguities. For this purpose, the desired ambiguity function of the waveform is plotted for all relevant range and Doppler shifts. Ideally, a single peak results at the origin of the plot, in which case the investigated function is unambiguous (cf. Figure 2.6). However, if further distinct peaks occur, the waveform is ambiguous.
On the other hand, it can be investigated how effects such as direct antenna crosstalk or signals from large targets impact the result. Figure 2.6 can be interpreted as direct antenna crosstalk, i.e., a portion of the non-delayed transmit signal that couples directly into the receiving branch. In addition to the expected peak at the origin, this direct coupling also produces an interference floor approximately 45 dB below the main peak. With the signal shape investigated here, a target signal should therefore be at most 40 dB below the antenna crosstalk in order to still be detected if we assume that a distance of 5 dB is sufficient for reliable detection.

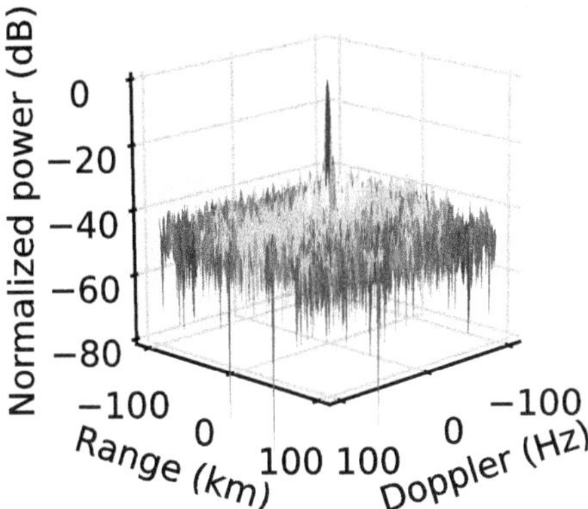


Figure 2.6.: Exemplary representation of an ambiguity function for an FM broadcasting station.

## 2.6. Beamforming

In order to make statements about the angle of arrival (AoA) of a signal with a coherently sampled array antenna, digital beamforming can be used. As boundary conditions, it is assumed that the target is sufficently far away that far-field conditions apply. This way, the incoming signal can be described in good approximation as a planar wavefront. It is further assumed that the wave travels through a homogeneous medium and that the propagation velocity is therefore constant. The following is a brief summary of the conventional beamformer described in [24, p. 23–32]. Figure 2.7 shows, using a two-dimensional example with a linear array, that the AoA leads to a time offset $\Delta\tau_i$ with which the wavefront reaches the individual antenna elements. To compensate for the influence of these delays, a linear, time-invariant filter function with the impulse response

$$h_n(\tau) = \frac{1}{N}\delta(\tau + \tau_n), \tag{2.30}$$

is defined. In this equation $1/N$ is a normalization factor with $N$ being the number of antenna elements and $\tau_n$ is the delay of the $n^{\text{th}}$ filter. The output of the beamformer in the time domain $y(t)$ can then be described by the convolution of this filter function with the signals of the individual antennas $(f(t, \vec{p}))$ [24, pp. 23–32]

$$y(t) = \sum_{n=0}^{N-1} \int_{-\infty}^{\infty} h_n(t-\tau) f_n(\tau, \vec{p}_n) d\tau. \tag{2.31}$$

Here $f_n(\tau, \vec{p}_n)$ describes the signal received at the $n^{\text{th}}$ antenna, depending on the antenna position $\vec{p}_n$.

The delay $\tau_n$ can be described with the unit vector on the propagation direction of the wavefront $\vec{a}$ and the propagation velocity in the medium $c$ as

$$\tau_n = \frac{\vec{a}\vec{p}_n}{c}. \tag{2.32}$$

Together with the definition of the wavevector

$$\vec{k} = \frac{\omega}{c}\vec{a}, \tag{2.33}$$

where $\omega = 2\pi f$ is the angular frequency, the delay can be described as a phase shift[6] with

$$\omega\tau_{\mathrm{n}} = \vec{k}^{\mathrm{T}}\vec{p}_n. \tag{2.34}$$

With this, the spatial properties of the array can be summarized in the array manifold vector

$$\vec{v}_{\mathrm{k}}(\vec{k}) = \begin{bmatrix} e^{-j\vec{k}^{\mathrm{T}}\vec{p}_0} \\ \vdots \\ e^{-j\vec{k}^{\mathrm{T}}\vec{p}_{(N-1)}} \end{bmatrix}. \tag{2.35}$$

Using the manifold vector the outputs of the array elements are defined as

$$\vec{f}(\omega) = f_{\mathrm{ref}}(\omega)\vec{v}_{\mathrm{k}}(\vec{k}) \tag{2.36}$$

using the reference signal $f_{\mathrm{ref}}(\omega)$, which would be received at the reference position of the array. To calculate the output of the beamformer $y(\omega)$ for a defined wavevector $\vec{k}_s$ of interest the beamforming vector

$$\vec{h}^{\mathrm{T}}(\omega) = \frac{1}{N}\vec{v}_{\mathrm{k}}^{H}(\vec{k}_s), \tag{2.37}$$

is obtained. Here $\cdot^{\mathrm{H}}$ is the conjugate transpose operator. This results in the equation for the conventional beamformer, also called delay-and-sum beamformer [24]:

$$y(\omega) = \vec{h}^{\mathrm{T}}(\omega)\vec{f}(\omega) \tag{2.38}$$

## 2.7. Array Modelling

The systems developed in this work use multiple antenna arrays. The influence of their geometry on the radiation pattern is described with the array factor $F(\theta, \varphi)$. According to [25], this factor is defined as

$$F(\theta, \varphi) = \sum_{n=0}^{N-1} \vec{h}_n e^{jk\vec{p}_n\vec{u}}. \tag{2.39}$$

[6] For narrowband signals, if the condition $B \cdot \Delta T_{\mathrm{max}} \ll 1$ is satisfied, phase shifts can be used instead of delays. Where $\Delta T_{\mathrm{max}}$ describes the maximum travel time of the incident wave between two array elements. [24, p. 34]

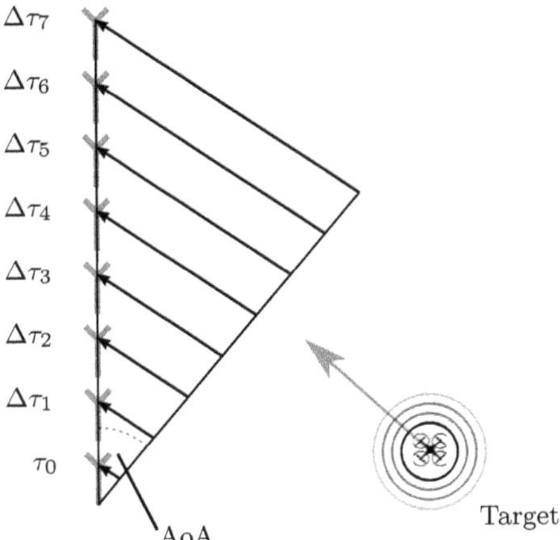


Figure 2.7.: Illustration of the AoA measurement principle

Where $\vec{h}_n$ is the $n^{\text{th}}$ element of the beamforming vector, $k = \frac{2\pi f}{c}$ is the wavenumber, $\vec{p}_n$ is the vector describing the position of the $n^{\text{th}}$ antenna element and $\vec{u}$ is the unit vector pointing towards the observed direction, it is defined as

$$\vec{u} = \begin{bmatrix} \sin(\theta)\cos(\varphi) \\ \sin(\theta)\sin(\varphi) \\ \cos(\theta) \end{bmatrix}. \tag{2.40}$$

The array factor is based on the assumption that all antennas are isotropic omnidirectional antennas. From the array factor, the directivity

$$D_{\text{F}}(\theta, \varphi) = \frac{4\pi |F(\theta, \varphi)|^2}{\int_0^{2\pi} \int_0^{\pi} |F(\hat{\theta}, \hat{\varphi})|^2 \sin(\hat{\theta}) \text{d}\hat{\theta} \text{d}\hat{\varphi}} \tag{2.41}$$

can be defined according to (6-102) from [26]. To model the behavior of the array with the actual antenna elements, the total directivity can be established with

$$D_{\text{tot}}(\theta, \varphi) = D_{\text{Ant}}(\theta, \varphi) D_{\text{F}}(\theta, \varphi), \tag{2.42}$$

assuming that the elements do not influence each other. A more detailed derivation of the array factor and its application can also be found in the thesis of F. Schwartau [27]. In this modeling, the effect of mutual coupling between the antenna elements is not yet taken into account. To get a good estimation of the array properties, this modeling is sufficient, but for a more accurate description, these coupling effects would have to be investigated more closely and modeled additionally.

## 2.8. Kalman Filter

A core component of every modern radar system is the tracker, which links past measurements of the target with the current ones. Trackers increase the accuracy of the data and offer the possibility to compensate for occasional missing measurement data. Such a tracker is also developed for the passive radar system described in this thesis. The crucial component of all these algorithms is usually a variaton of the Kalman filter (KF), which will be explained in more detail here. The KF named after Rudolf E. Kálmán was first described in the 1960s in [28]. It was invented to solve real-time linear filtering and prediction problems with numerical computation and was developed for discrete dynamic systems. The first application of an (extended) KF was in the Apollo guidance computer, the first embedded system ever used, and enabled the lunar landing by calculating the position and speed of the lander using the onboard inertial sensors and four earth-based Doppler radar systems [29]. Today, the KF is used in a vast range of applications, from position tracking (e.g. in aerospace and radar monitoring, or global navigation satellite system (GNSS)) to communication systems and data fusion applications. The broad applicability of this technique becomes clear if one considers the basic question of Kálmán's work:

> *"To have a concrete description or the type of problems to be studied, consider the following situation. We are given signal $x_1(t)$ and noise $x_2(t)$. Only the sum $y(t) = x_1(t) + x_2(t)$ can be observed. Suppose we have observed and know exactly the values of $y(t_0), ..., y(t)$. What can we infer from this knowledge in regard to the (unobservable) value of the signal at $t = t_1$, where $t_1$ may be less than, equal to, or greater than $t$? If $t_1 < t$, this is a data smoothing (interpolation) problem. If $t_1 = t$, this is called filtering. If $t_1 > t$, we have a prediction problem. Since our treatment will be general enough to include these and similar problems, we shall use hereafter the collective term estimation."* [28]

So the KF is a process, which can estimate the state of a given system. The basic principle of the KF is that the internal state of a system, described by its state vector $\vec{x}$ and the corresponding covariance matrix $\boldsymbol{P}$, is iteratively estimated for a time $t$. For this purpose, in each iteration, a system model is used to estimate the next system state (prediction phase). This predicted system state is then compared with measured values corresponding to the new time step (update phase). The result of this process is a weighted average of prediction and measurement. The basic steps of the iteration are depicted in Figure 2.8. To understand this process, it is important to realize that the process variables $\vec{x}$ and $\boldsymbol{P}$ are in three states within each iteration. The process starts with the original state, which is described here with the index $_0$. From this state within each iteration, a prediction to the time of the next measurement is made on the basis of a system model. This intermediate state is here marked by the index $_{\text{pre}}$. At the end of the iteration, the measured values that would theoretically result for the predicted (modeled) state are compared with the real measurement results. From the difference of these values, a corrective vector is then calculated to compensate for the error of the predicted state. The result of this correction is marked with the index $t$. Furthermore, with a KF, the state of the system does not need to be observed directly[7]. It is sufficient to measure quantities that depend on the hidden system state to be able to draw conclusions about it.

This chapter describes the basic equations of KFs and explains the aspects relevant for dimensioning. For a better understanding of the individual steps, the function is also explained using a simple example. For an exact formulation and a deeper understanding of the functions, the author recommends sources [29–31].

[7] The topic can also be described as Hidden Markov Model.

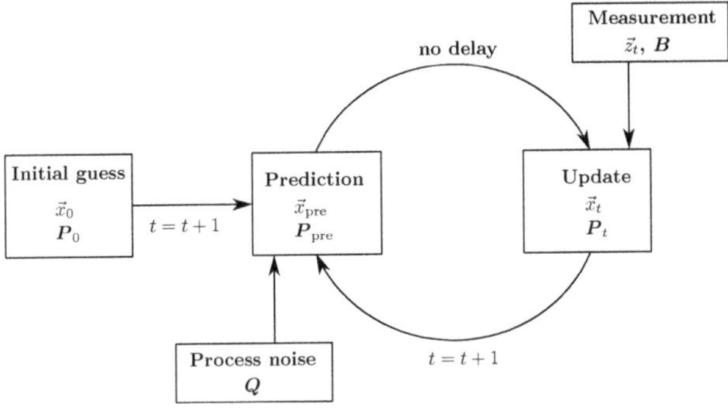


Figure 2.8.: Block diagram of the operation of a KF

## Basic Equations

The KF belongs to the class of Bayesian filters for estimating probability distributions. It is a special case for linear systems with normally distributed errors. It can be formulated with the five equations below [29]. A description of the process flow follows. For a better understanding of the operation, that is later used in the passive radar, a complete implementation is also explained using a simple example.

**Prediction**

$$\vec{x}_{\text{pre}} = \boldsymbol{F}\vec{x}_{t-1} + \boldsymbol{B}\vec{u}_{t-1} \tag{2.43}$$

$$\boldsymbol{P}_{\text{pre}} = \boldsymbol{F}\boldsymbol{P}_{t-1}\boldsymbol{F}^{\text{T}} + \boldsymbol{Q} \tag{2.44}$$

**Update**

$$\boldsymbol{K}_t = \boldsymbol{P}_{\text{pre}}\boldsymbol{H}_t^{\text{T}}\left(\boldsymbol{H}_t\boldsymbol{P}_{\text{pre}}\boldsymbol{H}_t^{\text{T}} + \boldsymbol{R}_t\right)^{-1} \tag{2.45}$$

$$\vec{x}_t = \vec{x}_{\text{pre}} + \boldsymbol{K}_t\left(\vec{z}_t - \boldsymbol{H}_t\vec{x}_{\text{pre}}\right) \tag{2.46}$$

$$\boldsymbol{P}_t = \left(\boldsymbol{I} - \boldsymbol{K}_t\boldsymbol{H}_t\right)\boldsymbol{P}_{\text{pre}} \tag{2.47}$$

| | |
|---|---|
| $\vec{x}$ | State vector |
| $\boldsymbol{F}$ | Transition matrix |
| $\boldsymbol{B}$ | Control matrix |
| $\vec{u}$ | Vector of deterministic distortions |
| $\boldsymbol{P}$ | Covariance matrix of the state |
| $\boldsymbol{Q}$ | Covariance matrix of the process noise |
| $\boldsymbol{K}$ | Kalman gain |
| $\boldsymbol{H}$ | Observation matrix |
| $\boldsymbol{R}$ | Covariance matrix of the measurements |
| $\vec{z}$ | Measurement vector |
| $\boldsymbol{I}$ | Identity matrix |

## Prediction Phase

The Kalman process always starts with the prediction phase. During initialization, an initial guess of the system state $\vec{x}$ and its covariance matrix $\boldsymbol{P}$ at time $t = 0$ is made. The KF then starts at the time $t = 1$ at which the first new measured value is recorded. Therefore, the first decision when designing a KF is the definition of the state vector $\vec{x}$. The state vector should contain not only the desired parameter but also derivatives of this parameter with respect to time so that a projection of the present state into the future is possible. For example, if the state is composed of a position $d$ and a velocity $v$, the new state would be predicted under the assumption of a constant velocity. This can be described with $d_{\text{pre}} = d_{t-1} + v_{t-1} \cdot \Delta t$ and $v_{\text{pre}} = v_{t-1}$. Where $\Delta t$ describes the variable time step between two measurements. This relationship can, of course, also be expressed with matrices

$$\boldsymbol{F}\vec{x} = \begin{bmatrix} 1 & \Delta t \\ 0 & 1 \end{bmatrix} \begin{bmatrix} d_{t-1} \\ v_{t-1} \end{bmatrix}.$$

The matrix described here, which models the state into the future, is called transition matrix $\boldsymbol{F}$. It results directly from the defined state vector and the physical system model, this is the first element of Equation (2.43). The next step is to model known deterministic influences on the system by adding a control term, which is the second element of Equation (2.43). This expression describes the known deterministic interference $\vec{u}$ of the system and maps it to the system state space using the control matrix $\boldsymbol{B}$. Such interference could be e.g. a known acceleration of the system or the steering angle of a car. The vector of influences $\vec{u}$ can contain any kind of data as long as a matrix $\boldsymbol{B}$ can be found, which maps it to the state. Thus all influences known to the observer are processed in Equation (2.43). However, a real system is not completely described by this equation. The influence of external parameters, which are unknown to the observer, is still missing. The real system at the time of measurement is therefore described by the following equation

$$\vec{x}_t = \boldsymbol{F}\vec{x}_{t-1} + \boldsymbol{B}\vec{u} + \boldsymbol{N}\vec{e}. \tag{2.48}$$

Here the unknown disturbances on the system are described in the process noise vector $\vec{e}$. Such influences could be e.g. unknown accelerations due to wind or any other effects, not described in the state and control terms. Analogous to $\vec{u}$ the process noise can have arbitrary elements[8] as long as they can be mapped to the state space with a matrix $\boldsymbol{N}$. Since the vector $\vec{e}$ is not known in the system, it cannot be considered in Equation (2.43). Instead, the contribution of $\vec{e}$ is described using the covariance matrix of the process noise $\boldsymbol{Q}$ and used in Equation (2.44), which performs the prediction of the covariance of the state. To understand the modeling of $\boldsymbol{Q}$, the first term of Equation (2.44) is of interest. This term describes the prediction of the state covariance matrix $\boldsymbol{P}$, which can be seen as the accuracy of the current state. This is also performed with the transition matrix $\boldsymbol{F}$ according to the general relation [32, p. 82]

$$\text{cov}(\boldsymbol{A}\vec{y}) = \boldsymbol{A}\text{cov}(\vec{y})\boldsymbol{A}^{\text{T}}. \tag{2.49}$$

With this in mind, the unknown errors represented by the vector $\vec{e}$ in Equation (2.44) can be modeled as a mean-free, normally distributed random function, which can be described by its covariance matrix $\boldsymbol{S} = \text{cov}(\vec{e})$. To get from this covariance matrix of $\vec{e}$ to the covariance matrix of the process noise $\boldsymbol{Q}$, the relation from Equation (2.49) is applied to the last term of Equation (2.48). This yields

$$\boldsymbol{Q} = \boldsymbol{N}\boldsymbol{S}\boldsymbol{N}^{\text{T}}. \tag{2.50}$$

[8] Often the process noise is defined in the format of the highest derivative contained in the state vector.

Comparing the Equations (2.43) and (2.44) it is noticeable that $\boldsymbol{Q}$ represents all uncertainties of the model between two sampling points. In a radar system, such uncertainties could be e.g. maneuvers of the target. As will be seen later in the example, the matrix $\boldsymbol{Q}$ largely determines the properties of the filter and must often be empirically adapted to the scenario. Too small values for this matrix result in a behavior in which the filter "trusts" its model. This suppresses the measurement noise very well, but the filter follows dynamic changes (whose origin is not described by $\boldsymbol{B}\vec{u}$) of the system only slowly, if at all. Values for $\boldsymbol{Q}$ that are too high, on the other hand, result in a strong following of the measured values and thus make the filter obsolete.

At the end of the prediction phase, the result is a predicted state $\vec{x}_{\text{pre}}$, which describes the new state of the system without the influence of the system noise and a prediction of the covariance matrix of the state $\boldsymbol{P}_{\text{pre}}$, in which the additional uncertainty by the system noise is represented.

## Update Phase

In the update phase, the predicted state is corrected with the help of actual measurements. To do this, three new terms are required in addition to the predicted status $\vec{x}_{\text{pre}}$ and the predicted covariance matrix of the state $\boldsymbol{P}_{\text{pre}}$. First of all, the vector with the measurements $\vec{z}$ must be defined. This vector can contain any number of arbitrary measurements as long as an observation matrix $\boldsymbol{H}$ can be found that maps the status vector $\vec{x}$ to the measurements $\vec{z}$ [9]. With these parameters, the three equations of the update phase can now be set up. First, the Kalman gain $\boldsymbol{K}$ is defined with Equation (2.45). $\boldsymbol{K}$ has two functions: on the one hand, it maps from the space of the measured values into the state space, and on the other hand, it weights the influence of the individual components. Next, Equation (2.46) is used to update the state. For the new $\vec{x}_t$ state, the predicted $\vec{x}_{\text{pre}}$ state is used as a basis and modified with a correction term. To get the correction term, first the error between the prediction and the measured values is calculated. For this purpose, the predicted status $\vec{x}_{\text{pre}}$ is mapped with $\boldsymbol{H}$ into the space of the measured values and then subtracted from the actual measured values. This leaves a vector with residuals, which in turn are mapped back into the state space with $\boldsymbol{K}$. This results in the correction vector, which is added to the prediction $\vec{x}_{\text{pre}}$.

The last step is to update the covariance matrix $\boldsymbol{P}$ with Equation (2.47). This records the improvement of the accuracy of the state by the new measurements. After the update, the process restarts with the prediction phase and a new set of measurements.

## Practical Example

For a better understanding of the KF a simple scenario is considered. Assuming an object in a one-dimensional space and neglected effects like friction, acceleration due to gravity or air resistance. The objective in this scenario is to track the position of the object with a very noisy position sensor.

[9] Here one of the advantages of the KF becomes apparent. It is sufficient to describe the mapping of the searched state $\vec{x}$ to the space of the measured values. The often more complex mapping from the space of measured values to the state space is not needed. This means that the KF is also capable of drawing results from an otherwise incomplete set of measured values and improving its estimation.

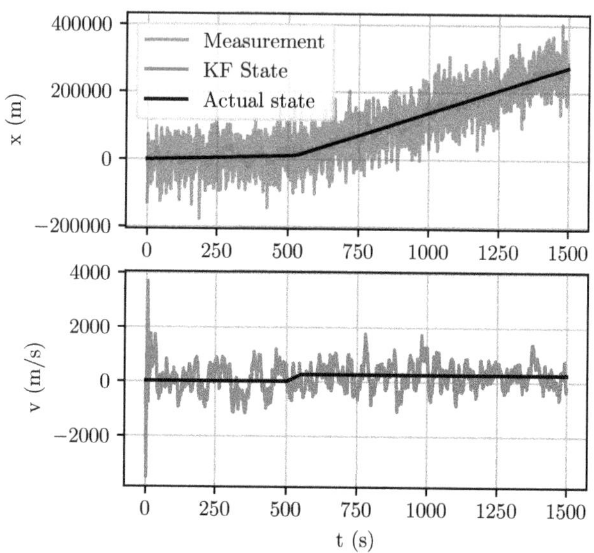


Figure 2.9.: Behavior of the KF with ideal $\sigma_{\mathrm{v_{error}}}$.

The object starts with a constant speed $v = 20\,\mathrm{m/s}$ at point $d = 0\,\mathrm{m}$. After a time of $t = 500\,\mathrm{s}$ the system is disturbed by accelerating the object for $50\,\mathrm{s}$ with $a = 5\,\mathrm{m/s^2}$. The defined scenario is shown in Figures 2.9 and 2.10 with black lines, the additional distortion is shown in Figure 2.11.

For the KF the state of the system is sampled with the simulated position sensor which yields the measured distance $d_{\mathrm{meas}}$. The time $\Delta t = 1\,\mathrm{s}$ is defined as the fixed sampling interval of the sensor and a normal distributed error with a variance of $\sigma^2_{\mathrm{d,meas}} = 50000\,\mathrm{m}^2$ is added to the measurement. Thus the measure vector $\vec{z} = \begin{bmatrix} d_{\mathrm{meas}} \end{bmatrix}$ and the corresponding covariance matrix $\boldsymbol{R} = \begin{bmatrix} \sigma^2_{\mathrm{d,meas}} \end{bmatrix}$ are obtained. The resulting measured values can be seen as blue lines in Figure 2.9. Next, the KF that is to be used to find the position of the object from the measurement data is defined. First of all the state vector and the transition matrix are defined. The state includes the current position of the object ($d$) and the derivation of this value over time, namely the velocity $v$. The state vector is $\vec{x} = \begin{bmatrix} d \\ v \end{bmatrix}$ with the corresponding covariance matrix[10] $\boldsymbol{P} = \begin{bmatrix} \sigma^2_{\mathrm{d}} & \sigma_{\mathrm{dv}} \\ \sigma_{\mathrm{vd}} & \sigma^2_{\mathrm{v}} \end{bmatrix}$. This shows an advantage of this type of filter; although only the position d is measured directly, information about the velocity is also obtained. Next, the matrix $\boldsymbol{F}$ must be selected so that it can transform a state at

[10] The covariance terms $\sigma_{\mathrm{dv}}$ and $\sigma_{\mathrm{vd}}$ are usually initialized with 0. The covariances in this example have the unit $\frac{\mathrm{m}^2}{\mathrm{s}}$. On the diagonal of the covariance matrix are the variances of the variables of the state vector. The variance is denoted as $\sigma^2$. The unit of $\sigma^2_{\mathrm{d}}$ is $\mathrm{m}^2$ and the unit of $\sigma^2_{\mathrm{v}}$ is $\frac{\mathrm{m}^2}{\mathrm{s}^2}$.

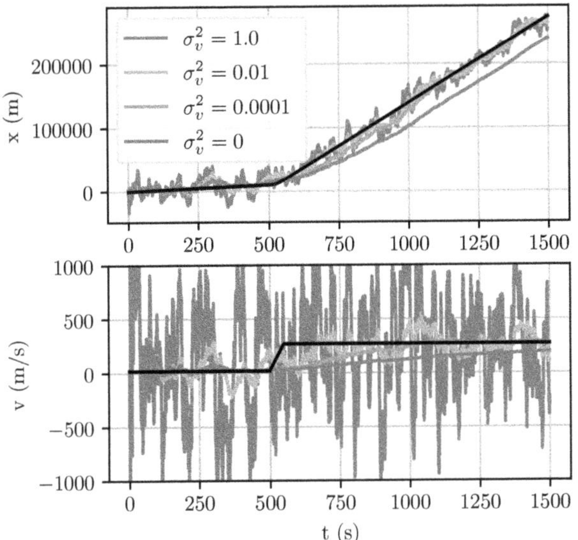


Figure 2.10.: Behavior of the KF with variations of $\sigma_{\mathrm{v}_{\mathrm{error}}}$.

time $t$ into a new state $t + \Delta t$. Since for the scenario $d_t = d_0 + vt$ and $v =$ const. apply, $\boldsymbol{F} = \begin{bmatrix} 1 & \Delta t \\ 0 & 1 \end{bmatrix}$ follows. The system does not contain any known deterministic disturbances, so $\vec{u}$ and $\boldsymbol{B}$ from Equation (2.43) can be omitted. Thus, all necessary components for the prediction phase of the KF are ready. Next $\vec{x}$ and $\boldsymbol{F}$ are applied to Equation (2.43)

$$\begin{aligned}
\vec{x}_{\mathrm{pre}} &= \boldsymbol{F}\vec{x}_{t-1} \\
\vec{x}_{\mathrm{pre}} &= \begin{bmatrix} 1 & \Delta t \\ 0 & 1 \end{bmatrix} \begin{bmatrix} d_{t-1} \\ v_{t-1} \end{bmatrix} \\
\vec{x}_{\mathrm{pre}} &= \begin{bmatrix} d_{t-1} + \Delta t \cdot v_{t-1} \\ v_{t-1} \end{bmatrix}.
\end{aligned}$$

With Equation (2.44) the prediction of the state covariance is expressed as

$$\begin{aligned}
\boldsymbol{P}_{\mathrm{pre}} &= \boldsymbol{F}\boldsymbol{P}_{t-1}\boldsymbol{F}^{\mathrm{T}} + \boldsymbol{Q} \\
\boldsymbol{P}_{\mathrm{pre}} &= \begin{bmatrix} 1 & \Delta t \\ 0 & 1 \end{bmatrix} \begin{bmatrix} \sigma^2_{\mathrm{d},t-1} & \sigma_{\mathrm{dv},t-1} \\ \sigma_{\mathrm{vd},t-1} & \sigma^2_{\mathrm{v},t-1} \end{bmatrix} \begin{bmatrix} 1 & 0 \\ \Delta t & 1 \end{bmatrix} + \boldsymbol{Q} \\
\boldsymbol{P}_{\mathrm{pre}} &= \begin{bmatrix} \sigma^2_{\mathrm{d},t-1} + \Delta t \sigma_{\mathrm{vd},t-1} & \sigma_{\mathrm{dv},t-1} + \Delta t \sigma^2_{\mathrm{v},t-1} \\ \sigma_{\mathrm{vd},t-1} & \sigma^2_{\mathrm{v},t-1} \end{bmatrix} \begin{bmatrix} 1 & 0 \\ \Delta t & 1 \end{bmatrix} + \boldsymbol{Q} \\
\boldsymbol{P}_{\mathrm{pre}} &= \begin{bmatrix} \Delta t^2 \sigma^2_{\mathrm{v},t-1} + \Delta t(\sigma_{\mathrm{vd},t-1} + \sigma_{\mathrm{dv},t-1}) + \sigma^2_{\mathrm{d},t-1} & \sigma_{\mathrm{dv},t-1} + \Delta t \sigma^2_{\mathrm{v},t-1} \\ \sigma_{\mathrm{vd},t-1} + \Delta t \sigma^2_{\mathrm{v},t-1} & \sigma^2_{\mathrm{v},t-1} \end{bmatrix} + \boldsymbol{Q}.
\end{aligned} \tag{2.51}$$

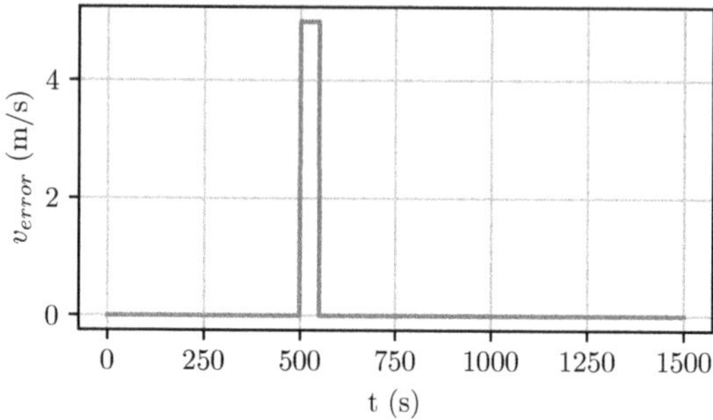


Figure 2.11.: Progression of the process noise over time.

The process noise from Equation (2.48) can be expressed as a speed error at the sampling time[11] which yields the process noise vector $\vec{e} = v_{\text{error}}$. The covariance matrix of $\vec{e}$ is denoted with $\boldsymbol{S} = \text{cov}\vec{e} = \begin{bmatrix}\sigma^2_{\text{v}_{\text{error}}}\end{bmatrix}$. According to Equation (2.48) the matrix $\boldsymbol{N}$ must have the property to map $\vec{e}$ into the state space. Since in this case $\vec{e}$ is a scalar velocity, $\boldsymbol{N} = \begin{bmatrix}\Delta t \\ 1\end{bmatrix}$ can be defined. Insertion into Equation (2.50) returns

$$
\begin{aligned}
\boldsymbol{Q} &= \begin{bmatrix}\Delta t \\ 1\end{bmatrix} \sigma^2_{\text{v}_{\text{error}}} \begin{bmatrix}\Delta t & 1\end{bmatrix} \\
\boldsymbol{Q} &= \begin{bmatrix}\Delta t^2 & \Delta t \\ \Delta t & 1\end{bmatrix} \sigma^2_{\text{v}_{\text{error}}}.
\end{aligned}
$$

Inserting $\boldsymbol{Q}$ into Equation (2.51) yields $\boldsymbol{P}_{\text{pre}} = \begin{bmatrix}\sigma^2_{\text{d,pre}} & \sigma_{\text{dv,pre}} \\ \sigma_{\text{vd,pre}} & \sigma^2_{\text{v,pre}}\end{bmatrix}$ with

$$
\begin{aligned}
\sigma^2_{\text{d,pre}} &= \Delta t^2(\sigma^2_{\text{v},t-1} + \sigma^2_{\text{v}_{\text{error}}}) + \Delta t(\sigma_{\text{vd},t-1} + \sigma_{\text{dv},t-1}) + \sigma^2_{\text{d},t-1} \\
\sigma_{\text{dv,pre}} &= \Delta t(\sigma^2_{\text{v},t-1} + \sigma^2_{\text{v}_{\text{error}}}) + \sigma_{\text{dv},t-1} \\
\sigma_{\text{vd,pre}} &= \Delta t(\sigma^2_{\text{v},t-1} + \sigma^2_{\text{v}_{\text{error}}}) + \sigma_{\text{vd},t-1} \\
\sigma^2_{\text{v,pre}} &= \sigma^2_{\text{v},t-1} + \sigma^2_{\text{v}_{\text{error}}}.
\end{aligned}
\tag{2.52}
$$

This completely describes the prediction phase of the KF. To formulate the update phase described in Equations (2.45) - (2.47) only the observation matrix $\boldsymbol{H}$ is missing. This matrix has the property to map from the state space into the space of the measured values. For the chosen example this relation is easy to formulate because a parameter from the state space is measured directly. It can be therefore be formulated as $H = \begin{bmatrix}1 & 0\end{bmatrix}$. With

[11] This is the most intuitive modelling. A modelling as partially constant or white noise is also conceivable. See [29, p.108].

this, the Kalman gain can be set up as follows:

$$\boldsymbol{K}_t = \begin{bmatrix} \sigma^2_{\text{d,pre}} & \sigma_{\text{dv,pre}} \\ \sigma_{\text{vd,pre}} & \sigma^2_{\text{v,pre}} \end{bmatrix} \begin{bmatrix} 1 \\ 0 \end{bmatrix} \left( \begin{bmatrix} 1 & 0 \end{bmatrix} \begin{bmatrix} \sigma^2_{\text{d,pre}} & \sigma_{\text{dv,pre}} \\ \sigma_{\text{vd,pre}} & \sigma^2_{\text{v,pre}} \end{bmatrix} \begin{bmatrix} 1 \\ 0 \end{bmatrix} + \sigma^2_{\text{d,meas}} \right)^{-1}$$

$$\boldsymbol{K}_t = \begin{bmatrix} \sigma^2_{\text{d,pre}} & \sigma_{\text{dv,pre}} \\ \sigma_{\text{vd,pre}} & \sigma^2_{\text{v,pre}} \end{bmatrix} \begin{bmatrix} 1 \\ 0 \end{bmatrix} \left( \sigma^2_{\text{d,pre}} + \sigma^2_{\text{d,meas}} \right)^{-1}$$

$$\boldsymbol{K}_t = \begin{bmatrix} \sigma^2_{\text{d,pre}} \\ \sigma_{\text{vd,pre}} \end{bmatrix} \frac{1}{\sigma^2_{\text{d,pre}} + \sigma^2_{\text{d,meas}}}.$$

$\boldsymbol{K}_t$ in this scenario is thus only dependent on the predicted variance of the position $\sigma^2_{\text{d,pre}}$, the predicted covariance between the velocity and the position $\sigma_{\text{vd,pre}}$, and the variance of the measured distance $\sigma^2_{\text{d,meas}}$. Next the equation for updating the status is determined with Equation (2.46):

$$\begin{aligned} \vec{x}_t &= \vec{x}_{\text{pre}} + \boldsymbol{K}_t \left( \vec{z}_t - \boldsymbol{H} \vec{x}_{\text{pre}} \right) \\ &= \begin{bmatrix} d_{\text{pre}} \\ v_{\text{pre}} \end{bmatrix} + \begin{bmatrix} \sigma^2_{\text{d,pre}} \\ \sigma_{\text{vd,pre}} \end{bmatrix} \frac{1}{\sigma^2_{\text{d,pre}} + \sigma^2_{\text{d,meas}}} \left( d_{meas} - \begin{bmatrix} 1 & 0 \end{bmatrix} \begin{bmatrix} d_{\text{pre}} \\ v_{\text{pre}} \end{bmatrix} \right) \\ &= \begin{bmatrix} d_{\text{pre}} \\ v_{\text{pre}} \end{bmatrix} + \begin{bmatrix} \sigma^2_{\text{d,pre}} \\ \sigma_{\text{vd,pre}} \end{bmatrix} \frac{1}{\sigma^2_{\text{d,pre}} + \sigma^2_{\text{d,meas}}} \left( d_{meas} - d_{\text{pre}} \right). \end{aligned}$$

To better understand the Kalman gain, this vector equation is split into two scalar equations, which are considered separately. The first equation describes the update of the position

$$d_t = d_{\text{pre}} + \underbrace{\frac{\sigma^2_{\text{d,pre}}}{\sigma^2_{\text{d,pre}} + \sigma^2_{\text{d,meas}}}}_{\text{Kalman gain}} \left( d_{meas} - d_{\text{pre}} \right),$$

The error to be corrected is set up as the difference between the measured and the predicted distance in the measurement space $(d_{meas} - d_{\text{pre}})$. This residual is then weighted by $\boldsymbol{K}_t$, which also maps it to the state space and adds it to the predicted distance. The example of the position $d_t$ shows how the weighting works. Assuming a perfect sensor $(\sigma^2_{\text{d,meas}} \rightarrow 0)$ the Kalman gain is 1, whereby the prediction is ignored and the measured value is used as the new position $(d_t = d_{meas})$. On the other hand, when approaching the other border case $\sigma^2_{\text{d,meas}} \rightarrow \infty$, the Kalman gain converges to 0. This leads to the measurement being discarded and only the prediction is adopted $(d_t = d_{\text{pre}})$. The same applies in principle to the update of the velocity

$$v_t = v_{\text{pre}} + \underbrace{\frac{\sigma_{\text{vd,pre}}}{\sigma^2_{\text{d,pre}} + \sigma^2_{\text{d,meas}}}}_{\text{Kalman gain}} \left( d_{meas} - d_{\text{pre}} \right).$$

In this case, however, the position error must be converted to a velocity. To better understand this mapping, the covariances from Equation (2.52) are used and again a perfect sensor with $(\sigma^2_{\text{d,meas}} \rightarrow 0)$ and no process noise $(\sigma^2_{\text{v}_{\text{error}}} \rightarrow 0)$ is assumed. This yields

$$v_t = v_{\text{pre}} + \frac{\Delta t \sigma^2_{\text{v},t-1} + \sigma_{\text{vd},t-1}}{\Delta t^2 \sigma^2_{\text{v},t-1} + \Delta t (\sigma_{\text{vd},t-1} + \sigma_{\text{dv},t-1}) + \sigma^2_{\text{d},t-1}} \left( d_{meas} - d_{\text{pre}} \right),$$

which generally describes the velocity update but does not illustrate the mechanism of the mapping clearly. For further simplification, the equation is considered at the time $t = 1$, after the covariance matrix has been initialized with[12] $(\sigma_{\mathrm{dv},0} = \sigma_{\mathrm{vd},0} = 0)$. The result is

$$v_t = v_{\mathrm{pre}} + \frac{\Delta t \sigma^2_{\mathrm{v},t-1}}{\Delta t^2 \sigma^2_{\mathrm{v},t-1} + \sigma^2_{\mathrm{d},t-1}} \left(d_{meas} - d_{\mathrm{pre}}\right).$$

The extreme cases of this equation illustrate how the determination of the derived variable from the measured values works. In the first case, the position at time $t = 0$ is exact and the speed is erroneous. For the variance of the position at time $t = 0$ the variance can be assumed with $\sigma^2_{\mathrm{d},0} \to 0$. A wrong speed in the state at $t = 0$ leads to a wrong prediction of the new position at $t = 1$. This position error is then corrected with the measured value as described above. The correction of the speed leads to

$$\begin{aligned} v_t &= v_{\mathrm{pre}} + \frac{\Delta t \sigma^2_{\mathrm{v},t-1}}{\Delta t^2 \sigma^2_{\mathrm{v},t-1}} \left(d_{meas} - d_{pre}\right) \\ &= v_{\mathrm{pre}} + \frac{d_{meas} - d_{\mathrm{pre}}}{\Delta t}, \end{aligned}$$

which exactly corresponds to the intuitive solution of the problem.

In the second case, if one turns the assumption around and assumes that the velocity is correct and the original position is erroneous, then the fraction in the Kalman gain approaches 0, resulting in

$$v_t = v_{\mathrm{pre}}.$$

Between these border cases, the KF weights in dependence of the variances $\sigma^2_{\mathrm{d},0}$ and $\sigma^2_{\mathrm{v},0}$. For subsequent time steps, the covariances $\sigma_{\mathrm{vd},t-1}$ and $\sigma_{\mathrm{dv},t-1}$ from the last time step are also added, so the covariance matrix does not only describe the current accuracy of the state, but also contains information about the history of the filter.

The last step of the Kalman iteration is the update of the covariance matrix $\boldsymbol{P}_t$ ,which is described by

$$\begin{aligned} \boldsymbol{P}_t &= \left( \begin{bmatrix} 1 & 0 \\ 0 & 1 \end{bmatrix} - \frac{1}{\sigma^2_{\mathrm{d,pre}} + \sigma^2_{\mathrm{d,meas}}} \begin{bmatrix} \sigma^2_{\mathrm{d,pre}} \\ \sigma_{\mathrm{vd},pre} \end{bmatrix} \begin{bmatrix} 1 & 0 \end{bmatrix} \right) \begin{bmatrix} \sigma^2_{\mathrm{d,pre}} & \sigma_{\mathrm{dv,pre}} \\ \sigma_{\mathrm{vd,pre}} & \sigma^2_{\mathrm{v,pre}} \end{bmatrix} \\ &= \begin{bmatrix} 1 - \frac{\sigma^2_{\mathrm{d,pre}}}{\sigma^2_{\mathrm{d,pre}} + \sigma^2_{\mathrm{d,meas}}} & 0 \\ -\frac{\sigma_{\mathrm{vd,pre}}}{\sigma^2_{\mathrm{d,pre}} + \sigma^2_{\mathrm{d,meas}}} & 1 \end{bmatrix} \begin{bmatrix} \sigma^2_{\mathrm{d,pre}} & \sigma_{\mathrm{dv,pre}} \\ \sigma_{\mathrm{vd,pre}} & \sigma^2_{\mathrm{v,pre}} \end{bmatrix} \\ &= \begin{bmatrix} \left(1 - \frac{\sigma^2_{\mathrm{d,pre}}}{\sigma^2_{\mathrm{d,pre}} + \sigma^2_{\mathrm{d,meas}}}\right) \sigma^2_{\mathrm{d,pre}} & \left(1 - \frac{\sigma^2_{\mathrm{d,pre}}}{\sigma^2_{\mathrm{d,pre}} + \sigma^2_{\mathrm{d,meas}}}\right) \sigma_{\mathrm{dv,pre}} \\ \sigma_{\mathrm{vd,pre}} - \frac{\sigma_{\mathrm{vd,pre}}}{\sigma^2_{\mathrm{d,pre}} + \sigma^2_{\mathrm{d,meas}}} \sigma^2_{\mathrm{d,pre}} & \sigma^2_{\mathrm{v,pre}} - \frac{\sigma_{\mathrm{vd,pre}}}{\sigma^2_{\mathrm{d,pre}} + \sigma^2_{\mathrm{d,meas}}} \sigma_{\mathrm{dv,pre}} \end{bmatrix} \\ &= \begin{bmatrix} \left(1 - \frac{\sigma^2_{\mathrm{d,pre}}}{\sigma^2_{\mathrm{d,pre}} + \sigma^2_{\mathrm{d,meas}}}\right) \sigma^2_{\mathrm{d,pre}} & \left(1 - \frac{\sigma^2_{\mathrm{d,pre}}}{\sigma^2_{\mathrm{d,pre}} + \sigma^2_{\mathrm{d,meas}}}\right) \sigma_{\mathrm{dv,pre}} \\ \left(1 - \frac{\sigma^2_{\mathrm{d,pre}}}{\sigma^2_{\mathrm{d,pre}} + \sigma^2_{\mathrm{d,meas}}}\right) \sigma_{\mathrm{vd,pre}} & \sigma^2_{\mathrm{v,pre}} - \frac{\sigma_{\mathrm{vd,pre}} \sigma_{\mathrm{dv,pre}}}{\sigma^2_{\mathrm{d,pre}} + \sigma^2_{\mathrm{d,meas}}} \end{bmatrix}. \end{aligned} \tag{2.53}$$

[12]The mapping works because the covariances $\sigma_{\mathrm{dv,pre}}$ and $\sigma_{\mathrm{vd,pre}}$ are calculated in the prediction phase and thus contain the relationship between the measured values and the state.

The variance of the position ($\sigma^2_{\mathrm{d},t}$) and the two covariance elements ($\sigma_{\mathrm{vd},t}$, $\sigma_{\mathrm{dv},t}$) are updated by scaling the predicted element with the term $\left(1 - \frac{\sigma^2_{\mathrm{d,pre}}}{\sigma^2_{\mathrm{d,pre}}+\sigma^2_{\mathrm{d,meas}}}\right)$. This term can take values between 0 and 1 depending on the variance of the measurement ($\sigma^2_{\mathrm{d,meas}}$). With ideal measurement values ($\sigma^2_{\mathrm{d,meas}} \to 0$) the predicted entries of the matrix would be scaled with the factor 0, which fits the behavior that was observed above, namely that a perfect measurement result also leads to an exact position determination. In the case of arbitrarily bad measured values ($\sigma^2_{\mathrm{d,meas}} \to \infty$) the fraction becomes zero and the total expression thus becomes one. The predicted values of the (co-)variances are thus adopted unchanged. For all cases between these two limiting cases, the values of the (co-)variances will be smaller in an update.[13] The update of the variance of the speed is described by

$$\sigma^2_{\mathrm{v},t} = \sigma^2_{\mathrm{v,pre}} - \frac{\sigma_{\mathrm{vd,pre}}\sigma_{\mathrm{dv,pre}}}{\sigma^2_{\mathrm{d,pre}} + \sigma^2_{\mathrm{d,meas}}}. \tag{2.54}$$

The border case for an arbitrarily bad measurement works the same way. For $\sigma^2_{\mathrm{d,meas}} \to \infty$ the fraction approaches zero and the predicted covariance is adopted unchanged. To get the behavior for the border case $\sigma^2_{\mathrm{d,meas}} \to 0$ the equations in Equation (2.52) are used. Under the assumption of a noise-free system $\sigma^2_{\mathrm{v_{error}}} = 0$, the following is obtained:

$$\sigma^2_{\mathrm{v},t} = \sigma^2_{\mathrm{v},t-1} - \frac{\Delta t^2\sigma^4_{\mathrm{v},t-1} + \Delta t\sigma^2_{\mathrm{v},t-1}\left(\sigma_{\mathrm{dv},t-1} + \sigma_{\mathrm{vd},t-1}\right) + \sigma_{\mathrm{dv},t-1}\sigma_{\mathrm{vd},t-1}}{\Delta t^2\sigma^2_{\mathrm{v},t-1} + \Delta t\left(\sigma_{\mathrm{dv},t-1} + \sigma_{\mathrm{vd},t-1}\right) + \sigma^2_{\mathrm{d},t-1}}.$$

Considering this equation again at time $t = 1$ and under the same conditions as the update of the position ($\sigma_{\mathrm{vd},t-1} = \sigma_{\mathrm{dv},t-1} = 0$) the equation further simplifies to

$$\begin{aligned}\sigma^2_{\mathrm{v},t} &= \sigma^2_{\mathrm{v},t-1} - \frac{\Delta t^2\sigma^4_{\mathrm{v},t-1}}{\Delta t^2\sigma^2_{\mathrm{v},t-1} + \sigma^2_{\mathrm{d},t-1}}\\ &= \left(1 - \frac{\Delta t^2\sigma^2_{\mathrm{v},t-1}}{\Delta t^2\sigma^2_{\mathrm{v},t-1} + \sigma^2_{\mathrm{d},t-1}}\right)\sigma^2_{\mathrm{v},t-1}.\end{aligned}$$

In the case where the initial position is correct ($\sigma^2_{\mathrm{d},t-1} \to 0$) and the velocity is incorrectly initialized, it was shown above that the velocity can be accurately corrected. In this case the update of the variance will be zero. Otherwise the value of the initial variance of position $\sigma^2_{\mathrm{d},t-1}$ controls the scaling of the velocity's variance $\sigma^2_{\mathrm{v},t-1}$.

Now that all the Kalman equations have been decomposed for this simple scenario, the importance of the added process noise becomes clear. It has been shown that the weighting between measurements and prediction depends largely on the entries of the covariance matrix $\boldsymbol{P}$. Equation (2.53) also shows that these entries either remain the same or become smaller during the update. Equation (2.52) shows that the entries of $\boldsymbol{P}$ increase over the product of time and $\sigma^2_v$, but $\sigma^2_v$ itself remains the same. This behavior would make sense in a noise-free system. The system would be updated with measured values until the position and speed are known with arbitrary precision, after which further corrections are unnecessary. $\sigma^2_v$ will therefore continue to approach 0 with each measured value and will react less and less to deviating measured values, no matter how far outside the prediction they may lie. To counteract this behavior it is important to define a suitable process noise.

[13] Any measurement is better then no measurement.

For the scenario described above, the behavior of the error can be described as in Figure 2.11. From this the variance of the process noise can be calculated [29, p. 123].

$$\sigma^2_{\mathrm{v}_{\mathrm{error}}} = \frac{\sum\limits_{n=1}^{N} v^2_{\mathrm{error}}(n)}{N} - \frac{\left(\sum\limits_{n=1}^{N} v_{\mathrm{error}}(n)\right)^2}{N}$$

For the example, with $N = 1500$ and $v_{\mathrm{error}} = \delta t \cdot a$, this yields $\sigma^2_{\mathrm{v}_{\mathrm{error}}} = 0.833\,\mathrm{m}^2/\mathrm{s}^2$. With this value and the above definitions for $\vec{x}$, $\boldsymbol{P}$, $\boldsymbol{F}$, $\boldsymbol{R}$, and $\boldsymbol{Q}$ a KF is now implemented in Python. For this purpose the implementation from the library filterpy[14] is used. The state is initialized with $\vec{x} = \begin{bmatrix} 2000\,\mathrm{m} \\ 5\,\mathrm{m/s} \end{bmatrix}$. The covariance matrix is initialized with $\boldsymbol{P} = \begin{bmatrix} 2000\,\mathrm{m}^2 & 0\,\mathrm{m}^2/\mathrm{s} \\ 0\,\mathrm{m}^2/\mathrm{s} & 1000\,\mathrm{m}^2/\mathrm{s}^2 \end{bmatrix}$. Figure 2.9 on page 22 shows the actual course of the position and velocity over time (black), the result of the KF (red) and the measured values (blue). The illustration shows how the output of the sensor can significantly be improved with the help of the KF. The result is a good compromise between averaging the measured values and following the disturbance.

If it is desired to change the filter behavior it is possible by adjusting the variance of the error $\sigma_{\mathrm{v}_{\mathrm{error}}}$. In reality, this term often has to be determined empirically, since it is usually not possible to determine it as exactly as in this simulated scenario.

Figure 2.10 presents the results of the KF for four variations. The red curve shows the already above predicted behavior for $\sigma_{\mathrm{v}_{\mathrm{error}}} = 0\,\mathrm{m}^2/\mathrm{s}^2$. Before the disturbance, the filter actually provides a good representation of the system and reflects the actual system state very well. After the disturbance the model changes only slowly and does not reach the actual system state. The larger $\sigma_{\mathrm{v}_{\mathrm{error}}}$ is chosen, the better the filter follows the disturbance and the worse the smoothing of the values becomes.

## 2.9. Non-linear Extension of the Kalman Filter

The filter originally defined by Kalman was intended for linear systems. In order to solve a wider range of problems with this technique, several extensions of the original approach were formulated to handle non-linear problems as well. Two of these approaches are presented in the following.

### 2.9.1. Extended Kalman Filter

The extended Kalman filter (EKF) represents the first implementation of a nonlinear KF. It was and still is used in many areas, e.g. state estimation in navigation systems, parameter estimation for training of neural networks and many other applications [33]. In principle, the EKF works similar to the KF, with the following differences. The transition matrix $\boldsymbol{F}$ and the control matrix $\boldsymbol{B}$ in Equation (2.55) are replaced by the nonlinear function $f(\vec{x}_{t-1}, u)$. This function is then linearized for the old state $\vec{x}_{t-1}$. The result is the Jacobian matrix in Equation (2.56). The prediction of the covariance matrix in Equation (2.57) is then again analogously to the KF. In the update phase the observation is described by the non-linear

[14] https://github.com/rlabbe/filterpy

function $h(\vec{x}_{\text{pre}})$. This is linearized in the predicted state $\vec{x}_{\text{pre}}$. The result is the Jacobian matrix described in Equation (2.58). The calculation of the Kalman gain $\boldsymbol{K}$ in Equation (2.59) and the update of the covariance matrix $\boldsymbol{P}$ in Equation (2.61) works again analogous to the KF. Equation (2.60) differs only in the calculation of the residual $\vec{z}_t - h(\vec{x}_{\text{pre}})$, where now the non-linear function $h$ is used. A detailed explanation of the EKF can be found in [34].

As with any linearization[15], the assumptions of the EKF work very well if the system is actually at the assumed operating point. The further the system moves away from this operating point, the worse its performance becomes. The distance of the system from the operating point can be caused by a bad initial guess or by too large time steps between updates[16]. A more detailed description of the problems with EKF can be found in [33, p. 35 ff.].

**Prediction**

$$\vec{x}_{\text{pre}} = f(\vec{x}_{t-1}, u) \tag{2.55}$$

$$\boldsymbol{F} = \left.\frac{\delta f}{\delta x}\right|_{x_{t-1},u} \tag{2.56}$$

$$\boldsymbol{P}_{\text{pre}} = \boldsymbol{F}\boldsymbol{P}_{t-1}\boldsymbol{F}^{\mathrm{T}} + \boldsymbol{Q} \tag{2.57}$$

**Update**

$$\boldsymbol{H} = \left.\frac{\delta h}{\delta x}\right|_{x_{\text{pre}}} \tag{2.58}$$

$$\boldsymbol{K}_t = \boldsymbol{P}_{\text{pre}}\boldsymbol{H}_t^{\mathrm{T}}\left(\boldsymbol{H}_t\boldsymbol{P}_{\text{pre}}\boldsymbol{H}_t^{\mathrm{T}} + \boldsymbol{R}_t\right)^{-1} \tag{2.59}$$

$$\vec{x}_t = \vec{x}_{\text{pre}} + \boldsymbol{K}_t\left(\vec{z}_t - h(\vec{x}_{\text{pre}})\right) \tag{2.60}$$

$$\boldsymbol{P}_t = (\boldsymbol{I} - \boldsymbol{K}_t\boldsymbol{H}_t)\,\boldsymbol{P}_{\text{pre}} \tag{2.61}$$

| | |
|---|---|
| $\vec{x}$ | State vector |
| $f$ | Non-linear transition function |
| $\boldsymbol{F}$ | Jacobian of the transition function |
| $\boldsymbol{B}$ | Control matrix |
| $\vec{u}$ | Vector of deterministic distortions |
| $\boldsymbol{P}$ | Covariance matrix of the state |
| $\boldsymbol{Q}$ | Covariance matrix of the process noise |
| $\boldsymbol{H}$ | Jacobian of the observation function |
| $h$ | Non-linear observation function |
| $\boldsymbol{K}$ | Kalman gain |
| $\boldsymbol{R}$ | Covariance matrix of the measurements |
| $\vec{z}$ | Measurement vector |
| $\boldsymbol{I}$ | Identity matrix |

[15] The linearization means a Taylor series broken off after the first order term. Thus this error grows the higher the degree of the original function is, or in other words the more "nonlinear" it is.

[16] In the case of strongly nonlinear functions, i.e. functions in which terms of the Taylor series with higher order have a strong influence, the distributions of the (normally distributed) random variables are also strongly distorted.

### 2.9.2. Unscented Kalman Filter - Sigma-Point Approach

An alternative to the EKF is the particle filter. This is a Monte Carlo approach in which the random distributions are modeled with a swarm of "particles". This swarm is then projected over the nonlinear mappings. In the target space of the mapping, the mean and variance of the mapped distribution can be determined. This method gives good results, even for nonlinear functions, and also has the advantage that no derivatives of the transition or observation function are needed. The disadvantage of this approach is that it is extremely computationally intensive and thus inefficient [33]. Therefore, this class of filters is not discussed in detail here.

The alternative to the EKF represents the class of sigma-point filters, which includes, among others, the unscented Kalman filter (UKF) presented in 1995 [35]. Instead of modeling the random distribution with a swarm of particles, it is modeled with a minimal set of weighted points - the sigma-points. For the selection of these points there is no clear rule and there are various approaches how to determine them. For this work the approach of the scaled unscented transform (SUT) described by Rudolph van der Merwe 2004 is used [33, p. 55 ff.] which is based on the work of Julier and Uhlmann [36]. In van der Merwes procedure, $2L+1$ weighted points are determined that describe the random distribution and its mean. Where $L$ corresponds to the dimension of the state vector. Figure 2.12 illustrates the problems with nonlinear mappings: The left column shows a strongly nonlinear mapping modeled with a Monte Carlo approach. It is clearly visible how the original Gaussian distribution is distorted by the mapping. In the middle, the linearized mapping as in the EKF is shown, which leads to large errors in both the mean and the covariance. Finally, the result of the sigma-point procedure is shown, which yields a good approximation of the actual mean and covariance.

The SUT uses three scaling parameters, $\alpha$,$\beta$,$\kappa$. On the choice of these points, van der Merwe writes:

> *"Choose the parameters $\kappa$, $\alpha$ and $\beta$. Choose $\kappa \geq 0$ to guarantee positive semi-definiteness of the covariance matrix. The specific value of kappa is not critical though, so a good default choice is $\kappa = 0$. Choose $0 \leq \alpha \leq 1$ and $\beta \geq 0$. $\alpha$ controls the "size" of the sigma-point distribution and should ideally be a small number to avoid sampling non-local effects when the nonlinearities are strong. Here "locality" is defined in terms on the probabilistic spread of $x$ as summarized by its covariance. $\beta$ is a non-negative weighting term which can be used to incorporate knowledge of the higher order moments of the distribution. For a Gaussian prior the optimal choice is $\beta = 2$ [37]*[Bibliography reference modified by author]. *This parameter can also be used to control the error in the kurtosis which affects the 'heaviness' of the tails of the posterior distribution."* [33, p.56]

To calculate the SUT, the auxiliary variable $\lambda$ is defined:

$$\lambda = \alpha^2 (L + \kappa) . \tag{2.62}$$

The first step of the SUT is the formulation of a set of weighted samples called sigma-points $\mathcal{S} = \{w_i, \mathcal{X}_i\}$, based on the original variables. Where $w$ is the weight of the sigma point and $\mathcal{X}$ is the value, or in the context of the Kalman filter, the state. The sigma-points are deterministically chosen to reflect the true mean and covariance of the original distribution described by $\vec{x}$ and $\boldsymbol{P}_{\mathrm{x}}$. The central $0^{\text{th}}$ sigma point $\mathcal{X}_0$ describes the mean value of the

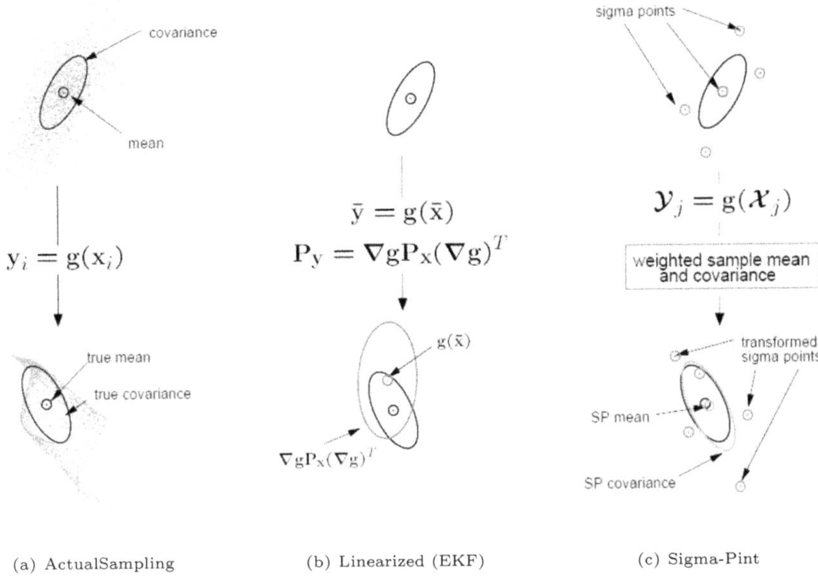


(a) ActualSampling (b) Linearized (EKF) (c) Sigma-Pint

Figure 2.12.: *"Demonstration of the accuracy of the scaled unscented transformation for mean and covariance propagation. a) actual, b) first-order linearization (EKF), c) SUT (sigma-point) A cloud of 5000 samples drawn from a Gaussian prior is propagated through an arbitrary highly nonlinear function and the true posterior sample mean and covariance are calculated. This reflects the truth as calculated by a Monte Carlo approach and is shown in the left plot. Next, the posterior random variable's statistics are calculated by a linearization approach as used in the EKF. The middle plot shows these results. The errors in both the mean and covariance as calculated by this "first-order" approximation is clearly visible. The plot on the right shows the results of the estimates calculated by the scaled unscented transformation. There is almost no bias error in the estimate of the mean and the estimated covariance is also much closer to the true covariance. The superior performance of the SUT approach is clearly evident."* Figure and caption cited from [33, p. 58]

old distribution function. The weights for the mean $w_0^{(m)}$ and the covariance matrix $w_0^{(c)}$ can be calculated with $\lambda$:

$$\begin{aligned} \mathcal{X}_0 &= \vec{x} \\ w_0^{(m)} &= 1 - \frac{L}{\lambda} \\ w_0^{(c)} &= 1 - \frac{L}{\lambda} + (1 - \alpha^2 + \beta) \end{aligned} \tag{2.63}$$

All other values, with their associated weights are then calculated as follows:

$$\begin{aligned} \mathcal{X}_i &= \vec{x} + \left(\sqrt{\lambda \boldsymbol{P}_{\mathrm{x}}}\right)_i & i &= 1, ..., L \\ \mathcal{X}_i &= \vec{x} - \left(\sqrt{\lambda \boldsymbol{P}_{\mathrm{x}}}\right)_i & i &= L+1, ..., 2L \\ w_i^{(m)} = w_i^{(c)} &= \frac{1}{2\lambda} & i &= 1, ..., 2L \end{aligned} \tag{2.64}$$

Where $\left(\sqrt{\lambda \boldsymbol{P}_{\mathrm{x}}}\right)_i$ is the $i^{\mathrm{th}}$ row or column of the matrix square root of $\lambda \boldsymbol{P}_{\mathrm{x}}$ [36]. The square root of the matrix is calculated with the Cholesky decomposition [30,33]. The sigma-points are then propagated through the non-linear function $f(\vec{x})$:

$$\mathcal{Y}_i = f(\mathcal{X}_i) \qquad i = 0, ..., 2L. \tag{2.65}$$

Finally, the new mean and covariance are calculated from $\mathcal{Y}$ with the pre-defined weights:

**Unscented Transformation**

$$\vec{y} = \sum_{i=0}^{2L} w_i^{(m)} \mathcal{Y}_i \tag{2.66}$$

$$\boldsymbol{P}_{yy} = \sum_{i=0}^{2L} w_i^{(c)} \left(\mathcal{Y}_i - \vec{y}\right) \left(\mathcal{Y}_i - \vec{y}\right)^{\mathrm{T}} \tag{2.67}$$

$$\boldsymbol{P}_{xy} = \sum_{i=0}^{2L} w_i^{(c)} \left(\mathcal{X}_i - \vec{x}\right) \left(\mathcal{Y}_i - \vec{y}\right)^{\mathrm{T}}. \tag{2.68}$$

With this transformation, the equations for the UKF can then be established [30,33]:

**Prediction**

$$\mathcal{X}_{\mathrm{pre}} = f(\mathcal{X}_{t-1}, \vec{u}) \tag{2.69}$$

$$\vec{x}_{\mathrm{pre}} = \sum_{i=0}^{2L} w_i^{(m)} \mathcal{X}_{pre,i} \tag{2.70}$$

$$\boldsymbol{P}_{\mathrm{pre}} = \sum_{i=0}^{2L} w_i^{(c)} \left(\mathcal{X}_{pre,i} - \vec{x}_{\mathrm{pre}}\right) \left(\mathcal{X}_{pre,i} - \vec{x}_{\mathrm{pre}}\right)^{\mathrm{T}} + \boldsymbol{Q} \tag{2.71}$$

**Update**

$$\mathcal{Z} = h(\mathcal{X}_{\text{pre}}) \tag{2.72}$$

$$\vec{z}_{\text{pre}} = \sum_{i=0}^{2L} w_i^{(m)} \mathcal{Z}_i \tag{2.73}$$

$$\vec{y} = \vec{z}_{t,meas} - \vec{z}_{\text{pre}} \tag{2.74}$$

$$\boldsymbol{P}_{zz} = \sum_{i=0}^{2L} (\mathcal{Z} - \vec{z}_{\text{pre}}) (\mathcal{Z} - \vec{z}_{\text{pre}})^{\mathrm{T}} \tag{2.75}$$

$$\boldsymbol{P}_{xz} = \sum_{i=0}^{2L} (\mathcal{X}_{\text{pre}} - \vec{x}_{\text{pre}}) (\mathcal{Z} - \vec{z}_{\text{pre}})^{\mathrm{T}} \tag{2.76}$$

$$\boldsymbol{K}_t = \boldsymbol{P}_{xz} (\boldsymbol{P}_{zz} + \boldsymbol{R}_t)^{-1} \tag{2.77}$$

$$\vec{x}_t = \vec{x}_{\text{pre}} + \boldsymbol{K}\vec{y} \tag{2.78}$$

$$\boldsymbol{P}_t = \boldsymbol{P}_{\text{pre}} - \boldsymbol{K}_t \boldsymbol{P}_{zz} \boldsymbol{K}_t^{\mathrm{T}} \tag{2.79}$$

| | |
|---|---|
| $\mathcal{X}$ | State of the sigma-points |
| $f(\mathcal{X}_{t-1}, \vec{u})$ | Transition function |
| $\vec{u}$ | Vector of deterministic distortions |
| $\vec{x}$ | State vector |
| $L$ | Dimension of $\vec{x}$ |
| $w_i^{(m)}$ | $i^{\text{th}}$ weight regarding the mean, corresponding to the $i^{\text{th}}$ sigma-point |
| $\boldsymbol{P}$ | Covariance matrix of the state |
| $w_i^{(c)}$ | $i^{\text{th}}$ weight regarding the covariance, corresponding to the $i^{\text{th}}$ sigma-point |
| $\boldsymbol{Q}$ | Covariance matrix of the process noise |
| $\mathcal{Z}$ | State of the sigma-points propagated through the observation function |
| $h(\mathcal{X}_{\text{pre}})$ | Observation function |
| $\vec{z}$ | Vector of the measurements |
| $\vec{y}$ | Residuum of the measurements and the predicted measurements |
| $\boldsymbol{P}_{zz}$ | Covariance of the predicted Measurement |
| $\boldsymbol{P}_{xz}$ | Cross-Covariance of the state and the predicted measurement |
| $\boldsymbol{K}$ | Kalman gain |
| $\boldsymbol{R}$ | Covariance matrix of the measurements |

Even though the processes of the UKF are more complex than for the linear KF, the dimensioning follows the same steps. Functions for transition $f$ and observation $h$ must be defined and process noise $Q$ must be modeled. In addition, the parameters $\alpha$, $\beta$ and $\kappa$ must be defined for the generation of the sigma points. The rest of the filter can be taken from predefined libraries.

# 3. Passive Coherent Receiver Hardware

The goal of this work is the development of a radio frequency (RF) passive detection system for airspace surveillance. The basis for detection is always the reception of radio signals in the microwave spectrum; however, depending on the origin of these signals, two modes of operation can still be distinguished:
On the one hand, the targets themself often carry an active transmitter, such as the telemetry transmitters of UAVs or the automatic dependent surveillance - broadcast (ADS-B) transmitters in commercial aircraft.
On the other hand, broadcasting transmitters can be utilized as radar illuminators. In this case, the reflection of these signals from the object is detected. This mode of operation is referred to as passive radar. This chapter deals with the development of a hardware platform that allows the investigation of both approaches for use in the detection of small aircraft / UAVs at maritime borders.

The first requirement to the system is therefore to offer the greatest possible flexibility. It should not be limited to individual radio systems, but should be able to work with the majority of usual radio and broadcasting systems. Therefore, a frequency range of up to 6 GHz should be usable. For the detection of targets, the possibility of determining the AoA of incoming signals is also relevant. The system shall therefore support beamforming as described in chapter 2.6. This means that a multi-channel coherent receiver is required and the system must support antenna arrays.

The first component in the receiving branch is the antenna. To support the wide frequency range and the different use cases the system should support different arrays and offer the possibility to switch quickly between them when the system is in operation. The system is supposed to be modular so that a later extension with further antenna systems for new tasks remains possible. For this purpose, controllable front-ends are developed, which offer the possibility to switch between different arrays and additionally provide the possibility to feed in a calibration signal.

The most important component of the system is the receiver itself, which has to be flexible as well. Therefore, the receiver is designed as a SDR; these have programmable front-ends and fast analog-to-digital converter (ADC), which can be used to mix down sections of the spectrum and digitize them for subsequent processing. The actual signal processing can then be computed digitally. An overview of the components required can be found as a block diagram in Figure 3.1.

This chapter introduces the individual components of this receiver system, as well as all the necessary procedures for operating it. The chapter is structured as follows; in Section 3.1 the modular core system and the necessary synchronization are described. This section has already been published at the "Sixth International Conference on Software Defined Systems" [38]. Subsequently, the antenna systems used are described in Section 3.2. Finally, in Section 3.3, measurements of the system's noise figure are shown. In addition, an

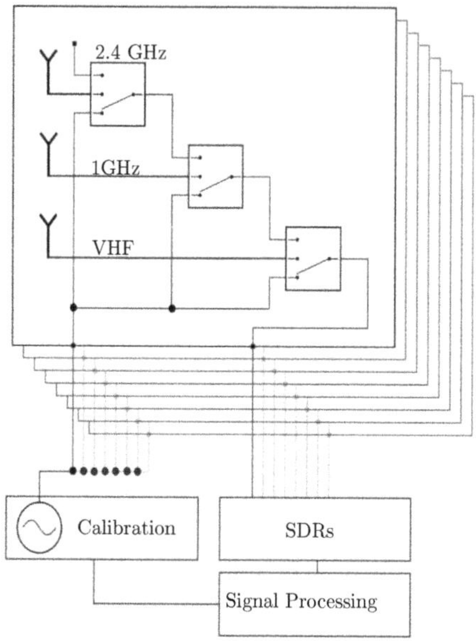


Figure 3.1.: Block diagram of the complete system

interesting consideration of how an alternative synchronization without the distribution of the LO signals would be possible can be found in Appendix B[1].

## 3.1. The Software-defined Platform

SDRs provide a great degree of freedom for various transceiver applications and are also used for multichannel applications like passive radar systems and direction-of-arrival (DoA) measurement setups [39, 40].
For such multichannel receiver applications, coherence between the individual channels is a fundamental requirement [41, 42].

To achieve coherence, most SDR systems are capable of synchronization with a common 10 MHz reference and some are also equipped with the means to share the LO signals directly. In this section the necessary means to combine multiple commercially-available software-defined radios to a multichannel coherent receiver system, using two Ettus Research USRP X310 SDRs, are described. A closer look at the mechanisms leading to phase ambiguities in systems synchronized to a common reference signal is provided. The cases in which this approach suffices for a phase-ambiguity-free setup and the cases in which further

[1]This was also part of the paper [38].

steps for the synchronization are necessary are described. These effects are also clarified with exemplary measurements.

Finally, a setup that ensures reliable synchronization of an arbitrary number of SDR systems is demonstrated. This setup is comprised of two Ettus X310 SDRs, combining four TwinRX daughterboards, resulting in an eight-channel system with a fixed initial phase relation. For this system long-term phase drift and a phase deviation on startup measurement are shown.

### 3.1.1. Hardware Description

A software-defined radio receiver is composed of an analog front-end and a subsequent analog-to-digital conversion. The analog front-end, which is basically a configurable analog receiver, converts a specified part of the spectrum down to a lower frequency band, where it can be converted to the digital domain. The digital receiver then performs baseband processing and, if necessary, demodulation by means of digital signal processing, which offers a great degree of flexibility.

The Ettus Research USRP X310 SDR features several ADCs and DACs as well as a Kintex 7 FPGA, which provides the platform's digital signal processing capabilities. Various high-speed interfaces can be used to offload the baseband data to a computer, which performs further data processing, and at the same time controls the SDR.

Each USRP is equipped with two slots for various analog front-ends, which are called daughterboards. One slot can handle a signal bandwidth of up to 160 MHz.

To allow basic synchronization between several USRP devices, the hardware is equipped with signal sources and distribution capabilities for a 10 MHz reference signal and a one pulse per second (1 PPS) signal. However, as will be seen in Section 3.1.2, synchronization by means of the 10 MHz reference signal does not ensure a well-defined phase relation between the individual channels.

The setup presented in this section combines two Ettus Research USRP X310 SDRs with a total of four TwinRX daughterboards to build an 8-channel coherent software-defined receiver. The Ettus Reseach TwinRX daughterboards support a frequency range from 10 MHz to 6 GHz and a bandwidth of up to 80 MHz per channel. Each channel is set up as a two-stage superheterodyne receiver.

### 3.1.2. Synchronization of the Analog Front-ends

The first and major task in building a coherent SDR setup is, similar to the setup of any coherent multichannel receiver, the synchronization of the analog receiver front-ends. To achieve this, the TwinRX architecture [43,44] is considered in more detail. Figure 3.2 shows a simplified block diagram of a single channel of a TwinRX daughterboard. It can be seen that the LO signals for the two mixers can either be generated by two internal oscillators or can be provided by the other daughterboard using the connectors depicted on the left. This option shall be referred to as LO sharing. A third option is to use internal switches (not shown in Figure 3.2) to feed one channel of a daughterboard with the LOs generated by the daughterboard's other channel.

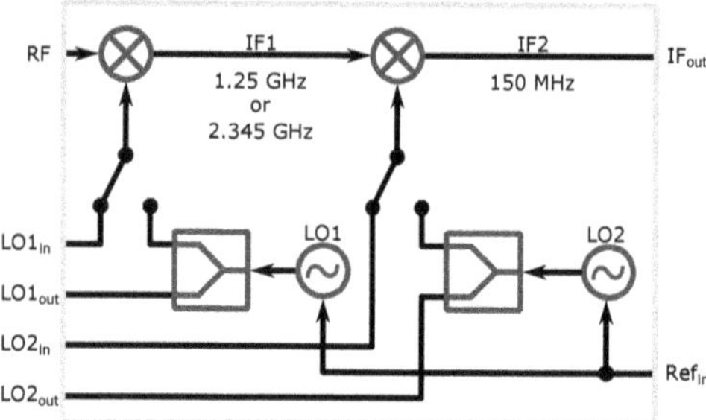


Figure 3.2.: Simplified block diagram of a single channel of a TwinRX daughterboard

#### LO sharing between daughterboards within one USRP

As already mentioned earlier, LO sharing allows to provide the four channels of the USRP's two daughterboards with the LOs generated by one of the system's channels. Thus, a single URSP equipped with two TwinRX daughterboards can be used as a 4-channel coherent receiver, which does not exhibit any phase ambiguity between the individual channels. LO sharing is well documented in [45].

#### Synchronization of the front-ends of more than one USRP

If more than four coherent receiver channels with a well-defined phase relation are needed, the LO signal, which is generated by one of the TwinRX daughterboards, has to be distributed using additional external hardware. To realize a system with 8 coherent receivers, the LO signals of one TwinRX channel is distributed using two 1:4 power dividers. Two additional amplifiers are used to mitigate the drop of signal power. A block diagram of the hardware setup is depicted in Figure 3.3.

Mini-Circuits ZN4PD-642W+ 4 Way 0 degree power splitters are used, which are specified for a frequency range from 1600 MHz to 6400 MHz and therefore cover all possible LO frequencies provided by the TwinRX daughterboards. The amplifiers used are Mini-Circuits ZX60-V83+ with a frequency range from 20 MHz to 4700 MHz and a typical gain of 15 dB.

### 3.1.3. Switchable LNA Front-ends

To achieve the modular, cascadable antenna subsystem, a controllable RF front-end is devised. The main component of the switchable LNA front-end is a 3:1 signal switch. One input is used as a link to the next antenna array, another one connects to a distributed calibration signal. The third input is used for the antenna input. On this port an additional low noise amplifier is installed. The amplifier input is protected with a bandpass filter to avoid a desensitization by interfering signals in other bands. The hardware consists of two printed circuit boards (PCBs), as can be seen in the block diagram in Figure 3.4. One PCB is a generic communication, control, and power supply unit, common for all front-ends. It assures a galvanic isolation to the rest of the system, produces the supply voltages for the RF board and handles the RS485 communication with the main system. The RS485

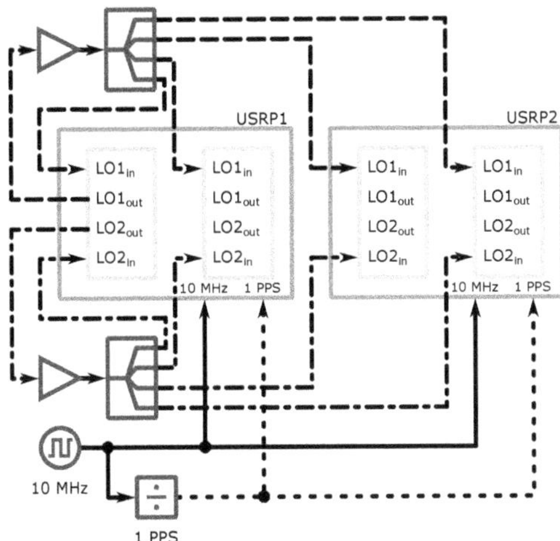


Figure 3.3.: Block diagram of the proposed LO and clock distribution architecture

bus together with the power supply is serially connected from each board to the next via a 4-wire interface. The received commands are transformed to control signals for the RF board. The second board houses the filter, low noise amplifier and RF switch. The amplifier is designed for a bandwidth between 50 MHz and 6 GHz. The input filter is interchangeable with any filter in a Mini-Circuits FV1206-4 case, which allows for customization of the RF board for different frequency bands. Each board has an unique 7-bit ID, which consists of four bits addressing the array and three bits to address each antenna individually. This leads to a maximum of sixteen arrays, comprised of up to eight antennas each, that can be used simultaneously.

### 3.1.4. Synchronization of the ADCs and FPGAs

Using the LO distribution setup described previously ensures that the received signals are coherent and exhibit a well defined phase relation at the interface between analog and digital domain. To realize a fully coherent receiver, it is furthermore necessary to ensure synchronization of the ADCs and the digital domain relative to the analog front-end.

The clock for the FPGAs and the ADCs is derived from a common 10 MHz reference signal, which can be fed into each USRP as shown in the simplified block diagram of the USRP depicted in Figure 3.5. In the setup, the 10 MHz reference signal is generated externally and distributed to both USRPs in order to achieve synchronous operation. Because the ADCs run at a fixed frequency of 200 MHz, no frequency ambiguity occurs between the ADCs, since the ADCs' operating frequency is an integer multiple of the 10 MHz reference signal. Additionally, the USRP provides a 1 PPS input, which is used to align the operations within the FPGAs. Once aligned, each FPGA has an internal counter, which can be used to synchronize commands like the begin of a recording or a switch in frequency.

The system's overall LO and clock distribution architecture is shown in Figure 3.3. The 10 MHz reference clock is generated by an IQD temperature-compensated crystal oscillator

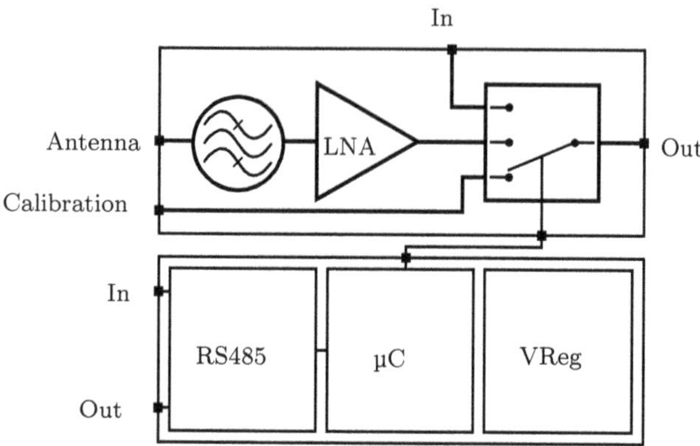


Figure 3.4.: Block diagram of the switchable LNA front-end

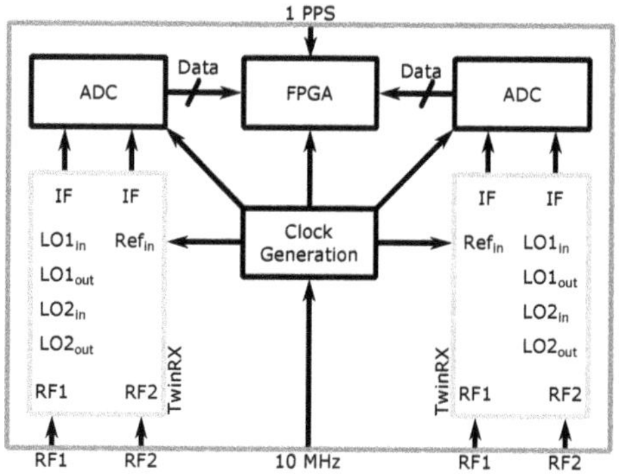


Figure 3.5.: Simplified block diagram of an Ettus Research USRP X310

(IQXT-220-1) and distributed using a Texas Instruments bus buffer IC (SN74LS125A). The 1 PPS signal is derived from this reference frequency using a microcontroller (ATTiny44) and also distributed using the SN74LS125A buffer IC.

### 3.1.5. Calibration

The clock distribution architecture introduced in Sections 3.1.2 and 3.1.4 allows the receiver's eight channels to operate in a fully coherent fashion with a well-defined phase relationship between the individual channels.

Nevertheless, there are remaining constant phase differences between the channels, which are mainly caused by length mismatches in the LO distribution network[2] and the connections to the individual antenna modules. In order to correct these phase differences a one-time calibration of the system is still necessary. Typically, this calibration is performed by recording a continuous-wave test signal, which is applied to each channel with a known phase offset[3]. Then, by aligning the phases of the signals received on the individual channels, the constant phase errors of the channels are determined.

### 3.1.6. Measurements

To verify the performance of the LO-sharing synchronization, several measurements are conducted. The measurements regarding the synchronization between the SDR channels are carried out using the Ettus X310 SDR system described above configured in the LO sharing synchronization scheme.

#### USRP phase deviation for multiple power-up cycles

Firstly, the phase deviation between the eight USRP channels is measured for 80 power-up cycles of the system using a common test signal at 2.45 GHz. For each start-up, the calibrated phase of the channels 2-8 is compared to the system's first channel, thus giving a total of $(8-1) \cdot 80 = 560$ data points for the phase deviation.

The distribution of the calibrated phase deviation of channels 2-8 is depicted in Figure 3.6(a). From this figure, it can be seen that the maximum phase deviation is below $\pm 0.8$ degrees, which shows that the receiver's synchronization works both accurately and reliably.

#### USRP phase stability over time

Additionally, the phase error is measured over a period of 60 minutes to verify phase stability over time. For this measurement, again a 2.45 GHz test signal is used. Figure 3.6(b) depicts the phase drift of the individual channels referred to the average phase of all 8 channels. The drift of all channels over the observed period of time is well below $\pm 0.3$ degrees, which is a very good result that again underlines the reliability of the system's synchronization.

## 3.2. Antennas

As mentioned above, the system can be extended modulary with up to 16 antenna arrays. The characteristics of the used arrays can be obtained by combining the characteristics of the individual antennas and the array factor as described in Section 2.7. So far, three different antenna arrays have been implemented. The first is a vertically polarized uniform circular array (UCA) for the band between 88 MHz and 137 MHz (very high frequency (VHF)). In

[2] Another problem with the LO distribution is that the mixers of the channel generating the LO can only be supplied with the LO using the channel's internal routing facilities rather than via the channel's input for external LOs. While in principle the latter approach should also be possible, the USRP firmware does not allow this option.

[3] This signal is generated for the respective frequency ranges and fed into the system via the RS-485 controlled LNA front-ends instead of the antenna signals. This also compensates for the constant phase errors caused by different cable lengths.

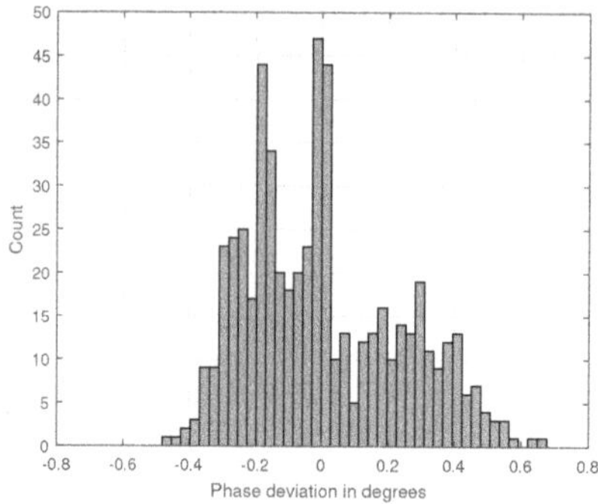


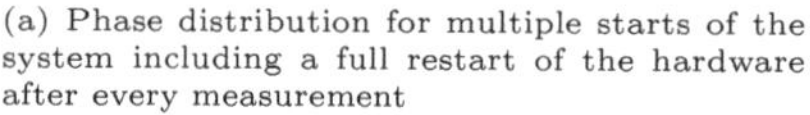
(a) Phase distribution for multiple starts of the system including a full restart of the hardware after every measurement

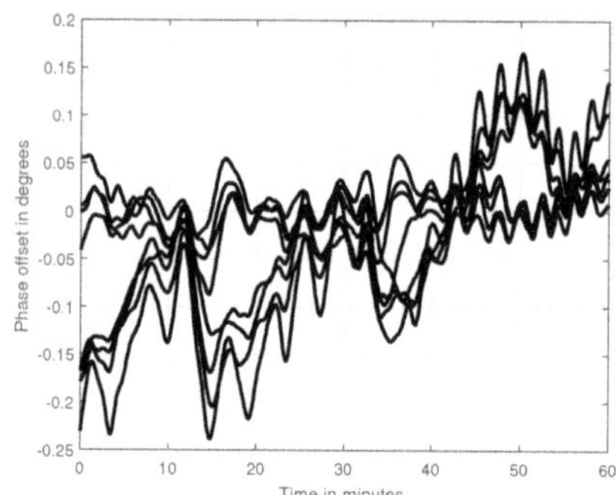


(b) Phase drift of the channels over time referenced to the average phase of all 8 channels

Figure 3.6.: Examinations of the phase stability of the setup.

this array, the eight antennas are distributed with a radius of $r = 96\,\mathrm{cm}$ [4], resulting in an element spacing of about $\frac{\lambda}{2}$. As antenna elements simple dipoles are used, their radiation characteristics can be described as [47]

$$\frac{S_r(\vartheta, \varphi)}{S_r(max)} = \frac{\cos^2(\frac{\pi}{2}\cos(\vartheta))}{\sin^2(\vartheta)} \tag{3.1}$$

and their gain can be assumed to be $G = 1.64$. Figure 3.7 shows the antenna patterns as a cut in the E and H planes for a beamforming in boresight ($\vartheta = \varphi = 0°$) for different frequencies. In this illustration, it is obvious that a clear, albeit broad, main lobe is formed. Opposite the main lobe a side lobe is formed, which is about 10 dB smaller than the main lobe. Between the main lobe and the side lobe are zeros in the pattern, of which the positions also shift slightly with the frequency. In elevation, these nulls are located at $\pm 90°$, i.e., directly above or, less relevant, below the antenna location. In azimuth they are approximately between $\pm 60°$ and $\pm 90°$ with respect to the main lobe.

The other two antennas are uniform linear arrays (ULAs). One of the arrays is designed for the 2.4 GHz industrial, scientific and medical (ISM) band. This array is built with patch antennas (see [26, p. 783 ff.]), with the individual elements spaced half a wavelength of the center frequency (6.2 cm) apart. The second ULA is to cover the frequency range from the 868 MHz ISM band to the range of ADS-B at 1090 MHz. Bow-tie antennas (see [48]) are used to cover this broad bandwidth. In addition, these antennas are placed in front of a reflector to shield one beam direction and thus minimize ambiguities in the bearing measurement. The radiation patterns of these two antenna types are simulated with Ansys HFSS. The input matching of the two antennas, as well as the E and H plane cuts of the whole array can be found in the appendix C.1. Figure 3.8(a) shows a model of the fabricated array antennas. Figure 3.8(b) shows the complete setup on a collapsible mast during a measurement campaign.

[4] According to [46] the half power beamwidth (HPBW) can be approximated as HPBW $\approx \frac{\lambda}{r} = 65.625°$.

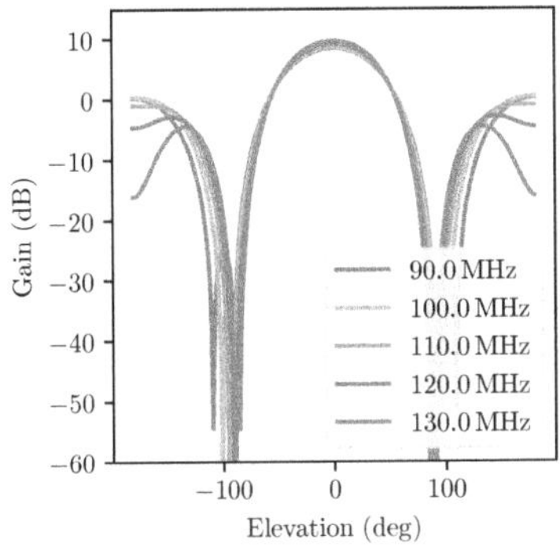


(a) Beam pattern in elevation

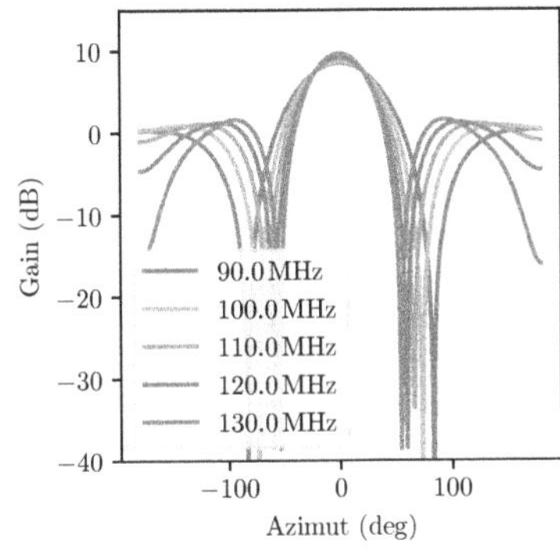


(b) Beam pattern in azimuth

Figure 3.7.: Characteristics of the UCA

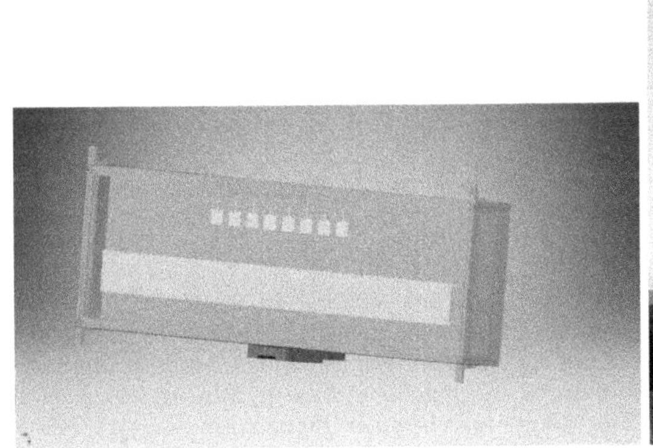

(a) The combined assembly for the two ULAs.

(b) VHF circular array with two linear arrays for 1 GHz and 2.4 GHz on top

Figure 3.8.: Representation of the antenna systems set up.

## 3.3. Noise Figure

When measuring weak signals, the noise characteristics of a system are crucial. These are described by the frequency-dependent noise figure $F$. $F$ indicates how much the SNR deteriorates due to influences of the system. The noise figure cannot be measured directly, however as described in [49, p. 501 ff.] it can be determined with the differential Y-factor method.

For this purpose, two measurements with a matched noise source are conducted. The source provides noise equivalent to two different temperatures ($T_1 > T_2$); these are used as the system's input signal. At the output of the system, the respective output noise powers $N_1$ and $N_2$ are determined. These are composed of the contributions from the noise source and the equivalent system noise temperature $T_S$ as follows:

$$Y = \frac{N_1}{N_2} = \frac{T_1 + T_S}{T_2 + T_S} > 1 \tag{3.2}$$

$$T_S = \frac{T_1 - YT_2}{Y - 1}. \tag{3.3}$$

The noise figure $F$ can be calculated via the relation

$$F = 1 + \frac{T_S}{T_0} \tag{3.4}$$

if both noise temperatures are known.
For the noise sources commonly used for these measurements [50], the excess noise ratio (ENR) is specified instead of the two temperatures. This is specified as follows

$$\mathrm{ENR} = \frac{T_1 - T_2}{T_0}. \tag{3.5}$$

For the lower temperature $T_2$, the noise source is operated in the switched off state. The noise temperature corresponds thereby to the actual temperature of the source. Therefore, in principle, the temperature of the noise source during the measurement plays a part. If it deviates significantly from $T_0 = 290\,\mathrm{K}$, this must be taken into account in the interpretation of the results. Since the measurements are performed in an air-conditioned laboratory, $T_0 = 290\,\mathrm{K}$ can be assumed with good approximation. Thus, with Equations (3.3), (3.4) and (3.5), the following equation for the noise figure can be established:

$$F = \frac{\mathrm{ENR}}{Y - 1} = \mathrm{ENR_{dB}} - 10\log_{10}(Y - 1) \tag{3.6}$$

The noise figure of the system is determined by this method for the 100 MHz and the 2.4 GHz band. For the measurements, the noise source was connected to the antenna inputs of the respective bands. The evaluation of the signal power was then performed on the basis of the digitized signals. The measured noise figure thus describes the entire system, including the analog-to-digital conversion. The results are shown in Figures 3.9 and 3.10. For the 100 MHz band, an average noise figure of 3.5 dB is obtained, and for the 2.4 GHz band, an average noise figure of 7.3 dB is obtained. In both measurements, interferences from external sources can also be observed, which have parasitically coupled into the system. In the 100 MHz band, these are local radio transmitters; the strong interference source in the 2.45 GHz band remains unknown. The noticeably deteriorated noise figure in comparison to the lower frequencies could also be the result of a broadband coupling of W-LAN signals.

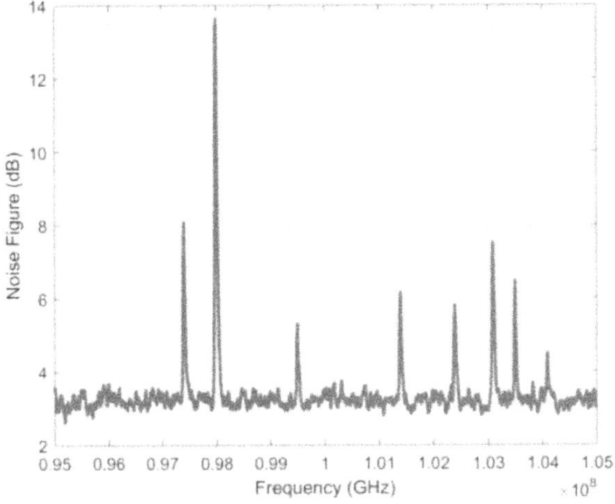


Figure 3.9.: The system noise figure measured for the VHF band. Parasitic coupled interference from local radio stations is clearly visible.

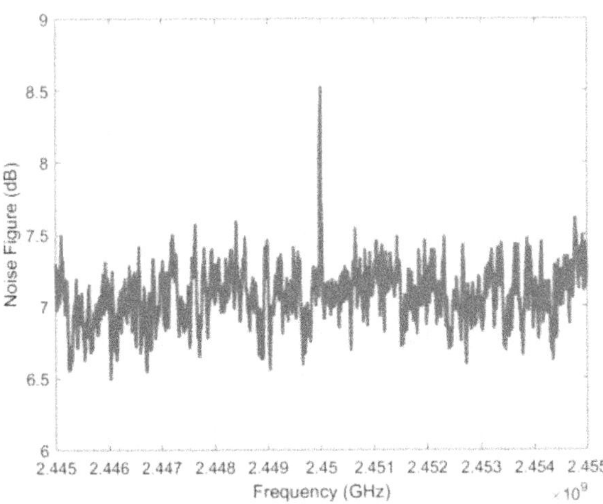


Figure 3.10.: The system noise figure measured for the 2.4 GHz band. A parasitically coupled carrier can be seen at 2.45 GHz, for which the source could not be determined.

## 3.4. Conclusion

In this chapter the hardware of a passive EM detection system based on an SDR platform was presented. For a coherent operation of such a system the synchronization of the single channels is essential. Therefore, the setup and synchronization of the 8 receivers was described and verified with measurements. Furthermore, a modular front-end was presented, which allows the extension of the system by up to 16 arrays with 8 antennas each. Currently, the system comprises of three such arrays, one UCA for the VHF band, and two ULAs for the 2.4 GHz ISM band, as well as for the 1 GHz band. Finally, noise figure measurements for the overall system are presented.

This system represents a versatile hardware platform that can be used for a wide variety of purposes. In the following chapters, a passive radar system based on this hardware will be described. In the Appendix D there is also an overview of the application of the system as a passive detection system for the detection of UAV.

# 4. Passive Radar Signal Processing

Passive radar systems were first introduced as early as World War II. The German radar system "Klein Heidelberg" used the transmitters of the British radar system "Chain Home" as an illuminator of opportunity [51]. As described in Chapter 2.4, the illuminator's position has to be known to interpret the measurements of a passive radar system. For high-power transmitters that operate from a fixed position, such as the "Chain Home" system, this information is, of course, known.
Modern passive radar systems no longer rely solely on enemy radar systems but use a variety of civilian radio services as illuminators. The variety ranges from broadcasting stations to mobile radio infrastructure and signals from GNSS satellites [52]. For some of these systems, especially those using smaller radio cells, it is sometimes a challenge to obtain reliable transmitter location information. The operator often has to rely on databases that are run by private individuals as a hobby (see [53]). These databases are admittedly remarkably comprehensive, but offer no guarantee for actuality or completeness. Especially for a mobile system, which has to be ready to use in different places quickly, it would be desirable to obtain this information in another way.
This chapter describes the required data processing of such a mobile passive radar system capable of determining illumination positions with cooperative targets, overcoming a crucial limitation of existing systems.

The mobile receiver system described in Chapter 3 is used as the basis of this system. In Section 4.1 the necessary preliminary considerations for selecting the transmitter system are described first. Subsequently, the individual data processing steps are described in Section 4.2. This description starts with the digitized samples of the eight single channels, describes the generation of the range-Doppler matrices, and ends with reducing the data to a list of range-Doppler data of potential targets. The complete signal processing is also shown as a block diagram in Figure 4.2. The processing method described here sets itself apart from the state of the art primarily by its ability to determine the illuminator position based on cooperative targets. This aspect will therefore be described in greater detail in Chapter 4.3. The fusion of the measurement data of the individual illuminators to target tracks is then finally shown in Chapter 4.4.

## 4.1. System Selection and Preliminary Considerations

The first step for the design of a passive radar system is to consider which illumination source should be used. The hardware described in Chapter 3 permits the use of many different systems. The selection of the system may thus be based on the intended application. As described in the Introduction, the task of the system is supposed to detect aircraft at long or medium distances. The most important requirement for the system is thus range and, therefore, high transmission power. According to Equation (2.21), the bandwidth of the system used also plays a role in the spatial resolution, but this is of secondary importance for the planned application. The planned system should also be mobile, i.e., as location-independent as possible. This results in the requirement of high coverage of the illuminator

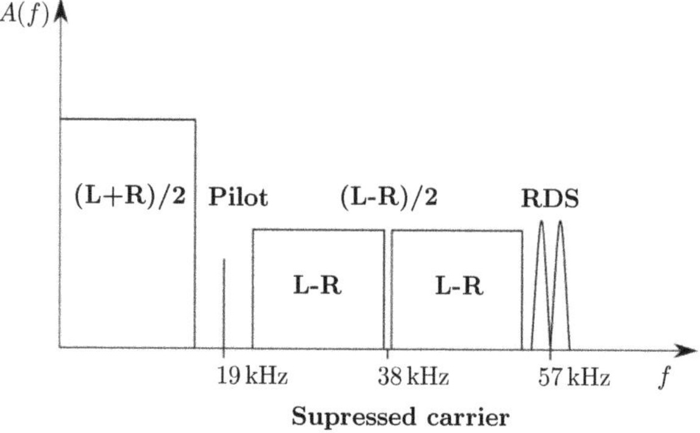


Figure 4.1.: Structure of the stereo multiplex signal spectrum [55, p. 858].

system. With these requirements, most bi-directional systems, such as mobile radio, are not suitable since they are usually organized in smaller cells. What remains are broadcast services, such as FM radio, DAB+, and DVB-T. These systems all have very good coverage and often work with higher transmitting powers to cover large areas with a few transmitter sites. The digital standards also have the advantage of higher and, above all, more constant bandwidths, which means that they can also provide very good spatial resolution. However, these standards are often operated as a single frequency network, which results in a higher complexity for the assignment of detections to single base stations. Therefore, FM radio is selected as the illuminator system of choice. However, a later addition of other systems, or possibly even mixed operation with different transmitter systems would be conceivable with the existing hardware. Technical details and information about the different radio standards can be found in the standards of the European Telecommunications Standards Institute (ETSI) [54], a good overview of the currently used broadcasting systems can be found in [55].

### 4.1.1. Transmitter Characteristics

The VHF FM radio system chosen as the illumination system is in use since 1940; a complete replacement with modern digital standards is intended, but it is not foreseeable when it will actually be implemented. The main reason for this is that analog receivers are still widely used. The system operates in the frequency range between 87.5 MHz and 108 MHz. It uses a channel spacing of 100 kHz, resp. 50 kHz in cable networks, with a desired channel spacing of 300 kHz by region-based frequency allocation.

Figure 4.1 shows the typical structure of the demodulated signal. The frequency-modulated sum signal of the two stereo channels lies in a bandwidth between 0 kHz and 15 kHz. This is followed by a pilot tone at 19 kHz, followed by the stereo difference signal amplitude-modulated at 38 kHz, using the 19 kHz as reference. It is located between 23 kHz and 53 kHz. Additionally, at 57 kHz data may be included using radio data system (RDS). As a further option, Subsidiary Communications Authorization (SCA) signals can be included in the channel between 60 kHz and 100 kHz [9, 55]. Thus, the maximum bandwidth of the signal can be estimated generously with 100 kHz. According to Equation (2.21), this results in a best possible resolution[1] of 1.5 km.

Of course, the actual achievable resolution depends on the effective bandwidth of the station and can thus vary for analog radio stations. In [56, 57], Griffith et al. describe this effect using the example of four radio stations in the London area. Resolutions between 1.8 km for a reggae station (effective bandwidth 83.5 kHz) and 16.5 km for a news station (effective bandwidth 9.1 kHz) are described. Due to this effect, the resolution will also vary with time. For example, it is common for many stations to interrupt the program with news at the top of the hour.

### 4.1.2. Separation of the Reference Signal

For the operation of the passive radar system, the separation of the illuminator signal from the target signals is an essential factor. The better the separation, the better the system will perform[2]. One approach to achieve this goal is to use transmitter polarization. For FM broadcasting, vertical, horizontal, and mixed (e.g. circular) polarization are specified. Which system is used depends largely on the terrain conditions; in flat terrain, with low transmitter height, vertical polarization is favorable, while in mountainous, forested terrain, horizontal polarization is preferred [58]. For transmitters using vertical or horizontal polarization, it is possible to separate the reference signal and surveillance channels employing polarimetric diversity [59]. To do this, the reference antenna has to be co-polarized with regard to the illuminator. The surveillance antennas are then operated cross-polarized.
Since the signals returned from geometrically complex targets also have cross-polarized components [60], the reference signal will be subject to a significantly more substantial attenuation than the target's signal. However, this effect will not work with mixed polarized transmitters. For the first tests with the passive radar, the reference channel is recorded with a separate, cross-polarized reference antenna to use polarimetric diversity. While this approach simplifies signal processing[3], it limits system functionality to linear polarized illuminators. Instead, a system is chosen that separates reference and surveillance channels via beamforming. The data processing described below and the results shown therefore refer to the system described in chapter 3, utilizing the circular VHF array.

## 4.2. Signal Processing

The complete process starting with the sampled signals of the individual antennas up to the tracked target is described. To provide an overview, the system is shown as a block diagram in Figure 4.2. In this section all steps from the "Complex Samples" to the "Target List" are described.

[1] Assuming that both targets are on the baseline and outside the space between transmitter and receiver ($\beta = \varphi = 0$).

[2] The importance of this aspect will be discussed in more detail later in Section 4.5.

[3] In general, the same techniques must be used for the signal processing when an additional reference antenna is used. The beamforming for the reference channel, however, can be omitted. On the other hand, the beamforming for the surveillance channel can be shifted to the end of the processing, cutting a large part of the parallel processing.

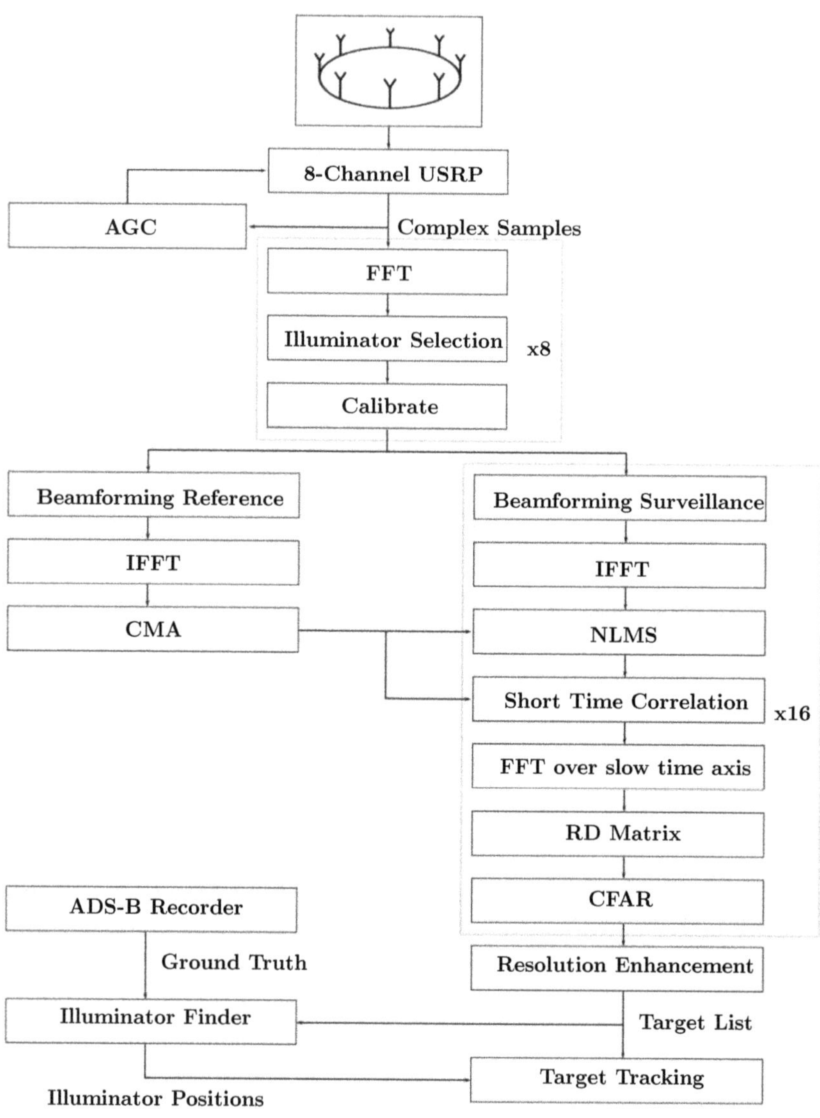


Figure 4.2.: Block diagram of the passive radar signal processing.

### 4.2.1. AGC

Before the actual data processing the automatic gain control (AGC) is implemented, which is responsible for making the best use of the receiver's dynamic range. In the system described here, this functionality is implemented with a simple digital control loop. In each pass of the signal processing, the root mean square (rms) power of the current signal block is compared to an ideal value. Based on the deviation from this value, the analog gain is then adjusted in 10 dB steps. Furthermore, a small hysteresis ensures a stable signal when the scenario is changing slightly. This type of gain control is comparatively simple, but proved to be sufficient during the field tests.

### 4.2.2. Illuminator Selection and Calibration

The first step in the data processing is to transform the eight individual antenna signals into the frequency domain using fast Fourier transformation (FFT). To save time, this step is calculated in parallel on a GPU using compute unified device architecture (CUDA) processing. The individual illuminator signals are then cut out of the spectrum.
When cutting out the spectrum in the frequency domain, several effects are obtained. On the one hand, the signal is ideally filtered and also mixed[4] down to the baseband. In addition, the signal is decimated in the frequency domain, which corresponds to a lowered effective sampling rate in the time domain [62,63]. For the measurements described in this work, the SDR is operated with a sampling rate of 25 MS/s and a total of $2^{24}$ complex samples (i.e., I and Q signal) per measurement are collected, which corresponds to a total measurement time of 671 ms. The Illuminator channels are then cut with a bandwidth of 200 kHz, which would corresponds to 134,217.7 samples and is rounded up to an even sample count of 134,218. This rounding results in the new sampling rate after decimation to be 200,000.41 Hz.

At the moment, the Illuminators are selected manually. However, it would also be conceivable to select the signals automatically, on the basis of their effective bandwidth, respectively their ambiguity function (see Section 2.5). Before further processing, for each transmitter, the eight channels are corrected in amplitude and phase by applying pre-recorded calibration data (see Section 3.1.5).

### 4.2.3. Beamforming

The system uses the delay-and-sum beamformer described in Chapter 2.6 to separate the reference signal from the surveillance channels. First, a beam is formed in the direction of the transmitter, which serves as the basis for the LOS reference signal. The direction of the transmitter is determined assuming that the LOS component is the largest signal component. Beamforming is calculated in 1° steps and the strongest component is assumed to be the reference. Additionally, 16 equidistant beams are formed for the surveillance

[4]When implementing a time domain mixer in the frequency domain, it must be noted that the mixing frequency can only be selected within the frequency resolution of the discrete fourier transform (DFT) [61], however, this limitation is not a drawback in this application. Furthermore, transforming the *rect* function from the frequency domain yields the *sinc* function in the time domain, which cannot be correctly represented with a finite number of samples (Gibbs phenomenon) [61]. If necessary, this can be counteracted with a suitable window function, applied in the frequency domain by means of a multiplication. However, due to the favorable distribution of the spectral components in the FM radio signal (described by Bessel functions [55]) this step could be omitted in this system.

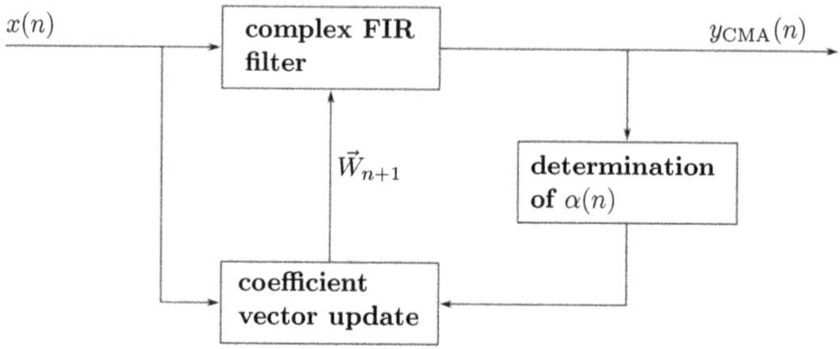


Figure 4.3.: General structure of the CMA as a block diagram [68].

channels.
The signal processing described in the following sections is based on this reference signal and the 16 surveillance channels.

### 4.2.4. Reference Signal Generation - CMA

As described above, the reference signal in this system is obtained by digital beamforming towards the transmitter. For the data processing, it is assumed that a "clean" reference signal is available. In other words, a LOS signal in an ideal additive white Gaussian noise (AWGN) channel that is affected only by free-space attenuation, range-induced time delay and noise. In a real channel, there will be additional delayed and scaled signal components due to multipath propagation.
The majority of these multipath components originate from reflections off buildings or terrain and are therefore static. These delayed signal components can lead to frequency-dependent, destructive interference at the receiver site, the so-called fading. For optimal operation of a passive radar system, these interferences must be compensated for in the reference signal [64].

For digital illuminators using complex QAM modulations, the multipath components can also reach the length of the transmitted symbols. In that case, additional intersymbol interference occurs, which leads to errors in the received signal. Such a signal must be reconstructed by using forward error correction techniques and then remodulated [65, 66]. For analog transmitting systems, like the FM broadcasting that has been chosen as the illuminator system, such procedures are not possible. Instead, for such mere frequency (phase) modulated signals, the constant envelope function of the signal can be utilized. For this purpose, an adaptive constant modulus algorithm (CMA) as described by Treichler and Agee [67] can be used. The implementation used here was devised by Benesty and Duhamel [68].

The CMA can be regarded as an adaptive FIR filter for equalization of the reference signal and it is based on the least mean square (LMS) algorithm. The general structure is shown as a block diagram in Figure 4.3. The output of the CMA $y_{\mathrm{CMA}}(n)$ is defined as a convolution of the signal with the filter's impulse response:

$$y_{\mathrm{CMA}}(n) = \vec{X}_n^{\mathrm{T}} * \vec{W}_n = \sum_{i=0}^{L-1} x(n-i)w_i. \tag{4.1}$$

Where $L$ is the length of the FIR filter, $\vec{X}_n$ is the vector of the last $L$ complex samples of the input $x(n)$ and $\vec{W}_n$ is the vector of $L$ complex weights, describing the filter. In the CMA algorithm, the filter coefficients for time $n+1$ are now obtained from the filter coefficients at time $n$ using a weighting function $\alpha(n)$ according to the LMS principle

$$\vec{W}_{n+1} = \vec{W}_n - \alpha(n)\vec{X}_n^* \tag{4.2}$$

where $*$ denotes the complex conjugation of $\vec{X}_n$. The difference to the LMS algorithm is now found in the formation of the weighting function

$$\alpha(n) = \mu_{\mathrm{CMA}} \left[|y_{\mathrm{CMA}}(n)|^2 - 1\right] y_{\mathrm{CMA}}(n). \tag{4.3}$$

This function does not compare the signal with an external reference, but again with the output function $y_{\mathrm{CMA}}(n)$. It exploits the known signal property of a constant amplitude and corrects the filter coeffients accordingly. The factor $\mu_{\mathrm{CMA}}$ describes a step size that influences the stability respectively the convergence speed of the filter.

For the system described here, the reference signal generated by beamforming is first transformed back to the time domain using an inverse fast Fourier transformation (IFFT) and normalized. The CMA is then implemented with a filter length of $L = 200$ and with a step size of $\mu_{\mathrm{CMA}} = 2e^{-4}$, which is emperically determined. The result of this process is used as the reference signal $\vec{S}_{\mathrm{Ref},n} = \vec{Y}_{\mathrm{CMA},n}$ in the further signal processing.

### 4.2.5. Surveillance Channel Generation - NLMS

After discussing the necessary measures for "cleaning" the reference signal in the previous chapter, this section considers similar procedures for the surveillance channels. The consideration begins with the question which signal components are contained in the individual surveillance channels. First of all there are the actual target signals. In addition to these sought-after signal components, there are also thermal noise components within the signal, and interference components from the direct illuminator signal, as well as multipath components of this signal.
For the moment, an ideal signal is assumed, which is only affected by noise. In this scenario, the ratio of the signal components can be calculated according to the radar equation (see Equation (2.19)). Furthermore, according to Equation (2.18) it can be assumed that the signals of the target components undergo a Doppler shift with respect to the reference signal[5]. However, in addition to these desired components, there are also parts of the reference LOS signal, as well as multipath components originating from reflections at stationary targets in the signal. The second component can be understood as a kind of unavoidable

[5] For the system, aircraft with a speed $> 80\,\mathrm{km/h}$ are assumed as the relevant targets. Targets moving tangential to the scenario do not have a Doppler component, which is another reason why it is important to work with as many different transmitter positions as possible.

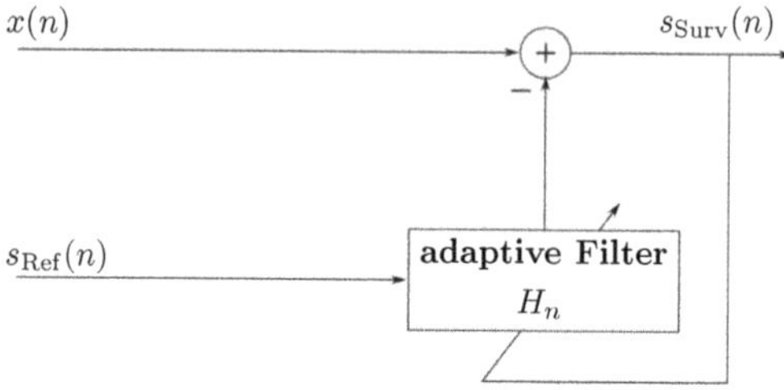


Figure 4.4.: Block diagram of the NLMS algorithm [13].

clutter.
Both of these unwanted signals can be described as phase-shifted but frequency-equal versions of the reference signal. Therefore, it is possible to find an FIR filter function that reproduces these signal components from the "cleaned" reference signal [13, 69, 70], thus making it possible to suppress them. The function for this filter must be calculated adaptively. A common method to calculate it is the normalized least mean square (NLMS) algorithm.

Figure 4.4 shows a block diagram of the procedure. The $n^{\mathrm{th}}$ output sample $s_{\mathrm{Surv}}(n)$ results from

$$s_{\mathrm{Surv}}(n) = x(n) - \vec{H}_n^{\mathrm{H}} \vec{S}_{\mathrm{Ref},n} = x(n) - \sum_{i=0}^{K-1} s_{\mathrm{Ref}}(n-i) h_i^*, \tag{4.4}$$

where $x(n)$ is the $n^{\mathrm{th}}$ sample of the beamformed surveillance signal $\vec{X}$, $\vec{H}$ is the weight vector of the adaptive FIR filter,

$$\vec{S}_{\mathrm{Ref}} = \begin{bmatrix} s_{\mathrm{Ref}}(n) \\ s_{\mathrm{Ref}}(n-1) \\ \vdots \\ s_{\mathrm{Ref}}(n-K+1) \end{bmatrix}$$

is the reference signal, generated by the CMA, $K$ is the filter length, and $h_i$ is the $i^{\mathrm{th}}$ weight from $\vec{H}$. The weights are calculated with

$$\vec{H}_{n+1} = \vec{H}_n + \mu_{\mathrm{NLMS}} \left[ s_{\mathrm{Surv}}(n) \vec{S}^*_{\mathrm{Ref}} \right] = \vec{H}_n + \mu_{\mathrm{NLMS}} \left[ x(n) \vec{S}^*_{\mathrm{Ref}} - ||\vec{S}_{\mathrm{Ref}}||^2 \vec{H}_n \right]. \tag{4.5}$$

Up to this point, the process is similar to that of a conventional adaptive LMS filter [71]. The difference in the NLMS algorithm is that the step size $\mu_{\mathrm{NLMS}}$ is dynamically adjusted, and is used to normalize the signal, resulting in

$$\mu_{\mathrm{NLMS}} = \frac{K \mu_0}{||\vec{S}_{\mathrm{Ref}}||^2}, \tag{4.6}$$

where $\mu_0$ is kept as a constant control parameter for the behavior of the algorithm. With this function for $\mu_{\text{NLMS}}$, Equation (4.5) can be rewritten to

$$\vec{H}_{n+1} = \vec{H}_n + \frac{K\mu_0}{||\vec{S}_{\text{Ref}}||^2}\left[s_{\text{Surv}}(n)\vec{S}^*_{\text{Ref}}\right] = \vec{H}_n + K\mu_0\left[\frac{x(n)\vec{S}^*_{\text{Ref}}}{||\vec{S}_{\text{Ref}}||^2} - \vec{H}_n\right]. \tag{4.7}$$

Like the CMA, this operation is also performed in the time domain. The cut spectrum of the surveillance channels is therefore transformed back into the time domain by means of an IFFT.
The results shown in this thesis have been obtained with a filter length of $L = 256$ and a step size of $\mu_0 = 0.02$. As shown in Figure 4.2, this algorithm had to be executed 16 times in parallel, so a fast, blockwise implementation is chosen. A more detailed description of this special implementation can be found in [72]. The result of this process is used as the surveillance signal $s_{\text{Surv}}(n)$ in the further signal processing.

### 4.2.6. Range-Doppler Matrix Generation

A (passive) radar system can generally find out two metrics about a target; the range and the frequency shift due to its velocity, or Doppler shift. These measurements are usually represented as a matrix, the so called range-Doppler matrix. It can be calculated by means of the cross-ambiguity function [13]

$$\psi_{m,k} = \sum_{n=0}^{N-1} s_{\text{Surv}}(n) \cdot s^*_{\text{Ref}}(n-m) \cdot \exp\left(\frac{-2\pi jkn}{N}\right). \tag{4.8}$$

The reference signal and the surveillance signal are evaluated for N samples, $m$ is an integer number of the range bin delay and $k$ is an integer number of the Doppler bin shift; the size of $N$ can also be used to select the integration time with $T = N/f_s$, where $f_s$ is the sampling frequency. A solution of Equation (4.8) is needed for all range and Doppler bins of interest, for all surveillance beams, and for all illuminators. To minimize the computational effort, a slightly simplified "batch algorithm" is usually used instead of the direct implementation of Equation (4.8) [73].
In this method, the signal is first divided into $Q$ blocks, each $P$ samples long, thus Equation (4.8) can be written as

$$\psi(m,k) = \sum_{q=0}^{Q-1}\sum_{p=0}^{P-1} s_{\text{Surv}}(qP+p) \cdot s^*_{\text{Ref}}(qP+p-m) \cdot \exp\left(\frac{-2\pi jk(qP+p)}{N}\right). \tag{4.9}$$

Assuming that the phase change within a block of length $P$ is negligible, the exponential function describing the Doppler shift can be excluded from the inner sum, yielding

$$\psi(m,k) = \sum_{q=0}^{Q-1}\left\{\sum_{p=0}^{P-1} s_{\text{Surv}}(qP+p) \cdot s^*_{\text{Ref}}(qP+p-m)\cdot\right\}\exp\left(\frac{-2\pi jkq}{N}\right). \tag{4.10}$$

In this function the inner sum corresponds to the numerical cross-correlation (COR) over the individual blocks and the outer sum corresponds to the DFT, which can be efficiently implemented with the FFT. Thus, the process can be described as follows

$$\text{FFT}\left\{\text{COR}\left\{s_{\text{Surv}}(qP+p), s_{\text{Ref}}(qP+p)\right\}\right\}. \tag{4.11}$$

The results of the correlations give the "fast" time offset, i.e., the range information. These individual columns are entered into a matrix along a "slow" time axis. Over the columns of this matrix the FFT is calculated, from which the Doppler shift results. The process is shown in Figure 4.5.

The dimensioning of the block sizes $Q$ and $P$ depends on the expected targets and the used illuminator system, in this case analog FM broadcasting service[6].
As described above, the individual 100 kHz channels are cut and processed with $L = 134,218$ samples at a sampling rate of $f_s = 200,000.41$ Hz. The values $P$ and $Q$ must now be chosen in a way that the maximum expected Doppler frequencies can be unambiguously represented with the FFT. To determine the maximum expected Doppler frequency, a maximal velocity of 900 km/h is assumed for the targets. According to Equation (2.18), the maximum Doppler shift can be calculated to be

$$\mathfrak{D}_{\text{max}} = \pm \left( \frac{2 \cdot f_{\text{I}} \cdot v_{\text{max}}}{c} \right). \tag{4.12}$$

Thus, at the maximum illuminator frequency $f_{\text{max}} = 108$ MHz, the targeted interval for the Doppler frequency is $\pm 180$ Hz. From Equation (4.10) it can be seen that the effective sampling frequency going into the FFT is $f_{\text{s,FFT}} = f_{\text{s}}/Q$. Rearranged for $Q$, with the condition that $f_{\text{s,FFT}} \geq 360$ Hz, a maximum length of the correlation sequence $Q_{\text{max}} \approx 555$ can be calculated. The minimum length of the slow time axis $P$ can be calculated with $P_{\text{min}} = L/Q$, with the previous results $P_{\text{min}} \approx 242$ is obtained. For an efficient calculation of the FFT the next power of two is chosen, which results in $P = 256$.
The length of a correlation block is set to $Q = 516$ samples. The size of a range bin here depends exclusively on the sampling frequency, $\Delta\mathfrak{R} = \frac{c}{f_{\text{s}}}$. This results in $\Delta\mathfrak{R} \approx 1.5$ km.
To save computing time, the correlation is only calculated for a range up to 100 km, i.e., for a total of 66 range bins. The size of a Doppler bin is given by $\Delta\mathfrak{D} = \frac{f_s}{Q \cdot P}$ and is $\Delta\mathfrak{D} = 1.5141$ Hz. An example of a range-Doppler matrix can be seen in Figure 4.6.

A side effect of this data processing is an additional process gain, by collecting the integral power of the complete signal into a single bin. This process gain depends on the effective bandwidth of the signal $B_{\text{eff}}$ and the integration time $t_{\text{i}}$ [13, p. 147] and can be estimated with

$$G_{\text{process}} = B_{\text{eff}} t_{\text{i}}. \tag{4.13}$$

## 4.2.7. Target Detection - CFAR

To extract the targets from the range-Doppler matrices, a CFAR algorithm is used. Such methods have been used in radar signal processing since the 1960s [15]. The aim of the process is to ensure a constant, low false alarm rate and thus to reduce the number of false detections that result from noise or clutter, minimizing the probability of a "false alarm". There are several methods, which are well documented in the literature [75, 76].
For the most part, these methods are extensions of the cell-averaging constant false-alarm rate (CA-CFAR) method also used in this work. The approach for the CA-CFAR is to compare the power of the cell under test (CUT) with the average power in its immediate

[6]It should be mentioned here that FM radio as an analog technology offers advantages in unambiguity in the range-Doppler domain. With digital standards ambiguities can occur, for example, due to periodically repeated pilot sequences. See e.g. [9, 13, 74].

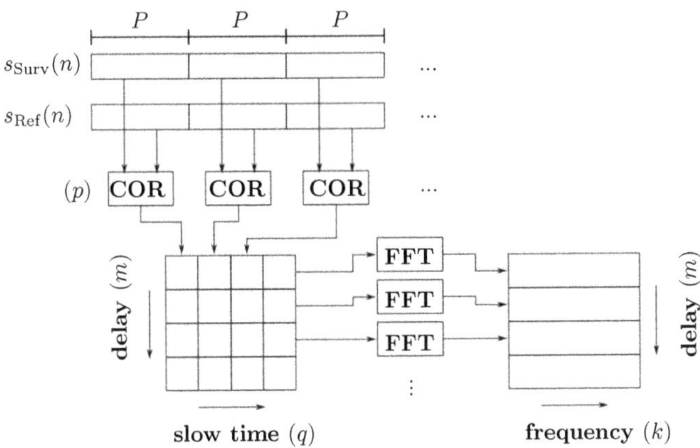


Figure 4.5.: Block diagram of the procedure used to calculate the range-Doppler matrices [13].

vicinity. A guard range is defined around the CUT, since it could still contain signal components of the actual target in case of strong reflections or leakage due to effects such as range migration[7]. For the results shown here, a guard range of 2 cells in range and Doppler is used. Outside the guard range a window is defined in which the surrounding signal power is averaged, for the results shown here a window size of 5 cells in range and Doppler around the CUT is used. For the target detection the power of the CUT is then checked to be above the average power of the window by a threshold value. For the results shown here, a threshold of 7.5 dB is used.
As a result of the CA-CFAR processing, a target list containing the range and Doppler information of the potential targets is obtained for each of the illuminators and each of the 16 surveillance angles An exemplary result is shown in Figure 4.7(a).

### 4.2.8. Interpolation and Target List Generation

At this point of signal processing, a list of potential targets obtained by the CA-CFAR exists for each illuminator and each of the 16 surveillance angles. These matrices are merged into a general target list. Next, under the assumption that due to the overlap of the surveillance beams, a real target should be visible in at least two beams, further filtering is performed. Any targets that appear in only one beam are considered artifacts and deleted. Next, it is assumed that targets in neighboring cells (range and Doppler) are actually associated with a single target. These neighboring cells are searched with a recursive algorithm and reduced to the cell closest to the mean of this cluster.

[7]Such effects occur, for example, when a target moves so fast that it is detected in multiple range-cells during a single acquisition.

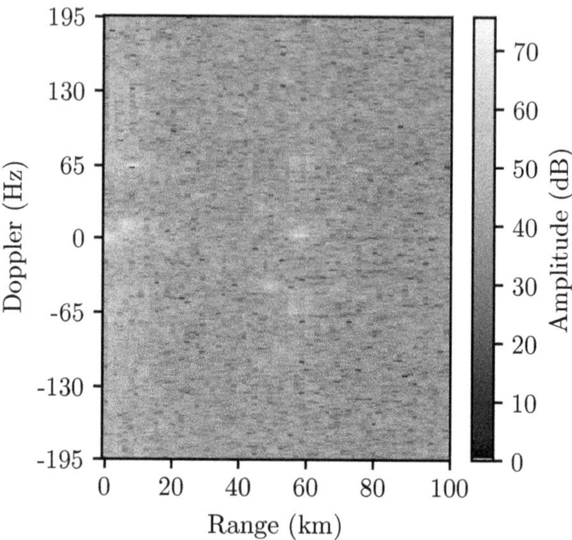


Figure 4.6.: Example of a range-Doppler matrix before interpolation.

All parameters of the targets that the system can determine are now already available, namely range, Doppler and the angle. So far the data has been processed with the expected resolution of the system in mind. The resolution describes how far apart two targets have to be in order to be represented as two separate targets. However, it does not give any measure of the achievable accuracy of the system for the individual targets. The accuracy with which any characteristic of a target can be determined can be described with a standard deviation. According to [15], the standard deviation of the measured values is

$$\sigma_M = \frac{k_\mathrm{M}}{\sqrt{\mathrm{SNR}}} \cdot \Delta M. \tag{4.14}$$

In this equation $k_\mathrm{M}$ describes a dimensionless constant that must be defined for the measurement method. $\Delta M$ describes the resolution of the measurement and the SNR is related to the current measurement.

This means that the accuracy of each measurement depends on the SNR of that measurement, so strong signals can be detected more accurately. In order to make a direct statement about the standard deviation, the factor $k_M$ is still missing for this system, which has not yet been determined for the final system.

However, to give an estimate of what to expect, the system can be compared with results from the literature: In [13], a system is described that in most aspects has similar properties

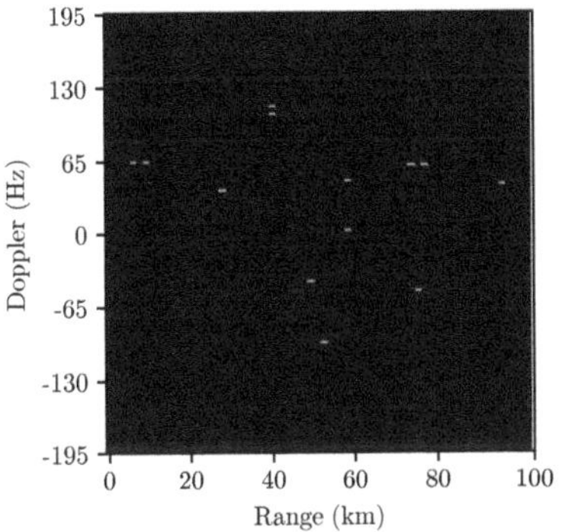


(a) Results of constant false-alarm rate (CFAR).

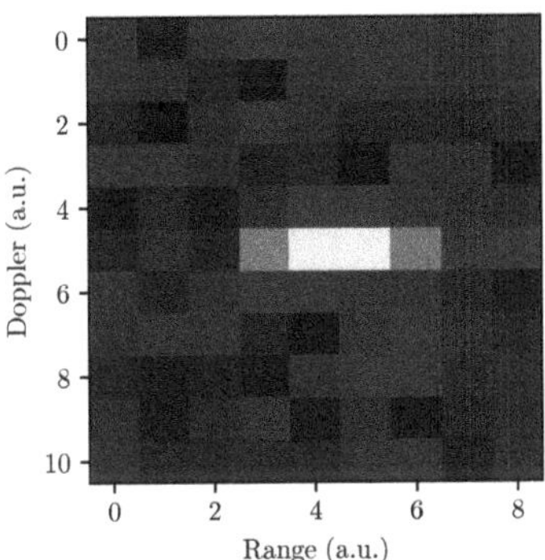


(b) A cutout of a target.

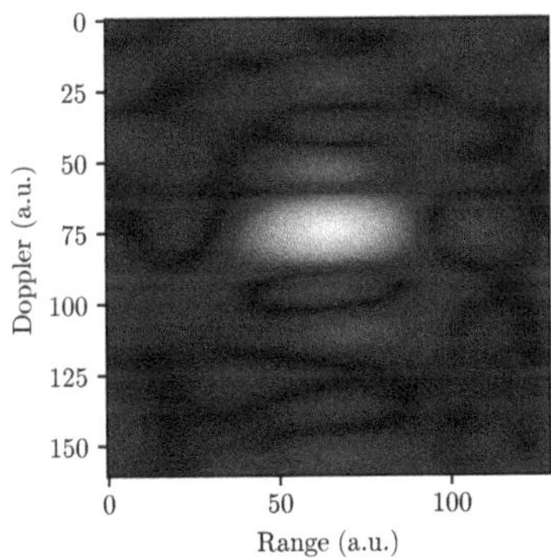


(c) Interpolation of the cutout section.

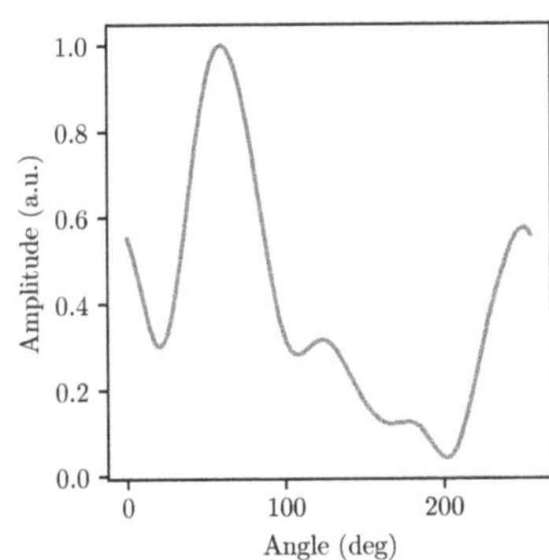


(d) Interpolation of the angle information.

Figure 4.7.: The steps in the processing of target information.

as the system used in this work. For this reference system, accuracies up to an order of magnitude better than the resolution are given for an SNR of 20 dB.

To improve the resolution of the system, a 2-dimensional sinc-interpolation [77] is performed on the data. In a first step, the range and Doppler axes are interpolated for each of the 16 surveillance angles. For each target an area of ±4 cells in range and ±5 cells in Doppler is cut out. This results in an 9x11 matrix (see Figure 4.7(b)).
The actual interpolation is then performed with a 2D FFT and zero padding. Afterwards the result is transformed back with the corresponding IFFT. The degree of interpolation can be defined by the length of the zero padding. In the results shown here, interpolation is always by a factor of 16 in range and Doppler. An example of an interpolated target matrix is shown in Figure 4.7(c).

After this operation has been performed for all 16 surveillance angles, the angle information is interpolated as well in the following step. The first task is to find the range-Doppler value along which the interpolation is to be performed. For this purpose, the interpolated range-Doppler matrices for a target are searched for the entry with the highest power. For the range-Doppler coordinate of this entry the amplitude over the 16 angles is taken as starting point for the interpolation. The angle information is then again interpolated by Fourier transformation, zero padding and inverse transformation.
Also for the angle data a scaling factor of 16 is chosen. An example is shown in Figure 4.7. The result has an angle bin size of 1.4°. After this step, for each transmitter a target list is available, which contains information about range, Doppler and an angle information.

## 4.3. Illuminator Locator

In the previous sections, a list of information about Range, Doppler and the angle to the target was compiled. In order to combine this data into usable information about the three-dimensional position of the target and its velocity vector, one crucial piece of information is still missing: the position of the transmitters. With the opportunistic use of illuminators in a passive radar system, this information is often not obtained easily. To address this drawback, a novel method for illuminator detection using cooperative targets is devised in this section. The results of this section have already been published in the IEEE Transactions on Aerospace and Electronic Systems [78] and are cited from this publication in the following. In parallel to this work, M. Malanowski has also been working on this technique and presented his results at the 2020 IEEE International Radar Conference [79] a few months before the submission of the paper presented here. The results of Malanowski were taken into account and included before submission.

### 4.3.1. Introduction

This work addresses the problem of finding illuminators of opportunity for a versatile, mobile passive radar system. In such a bistatic system, information on the transmitter positions is essential for the interpretation of the measured range-Doppler data. In most passive radar systems described in the literature the positions of these non-cooperative[8] illuminators are assumed to be known [80–82]. Various databases exist, from which such information can be retrieved, but required data such as location, height, transmitter frequencies and power may have to be compiled from different sources. Such public databases

[8]Non-cooperative in this context means that the transmit signal is not directly provided.

are often organized as private projects and therefore do not guarantee the actuality, accuracy or completeness of the data they contain. Furthermore, commercial use is often not permitted. It would therefore be desirable to use a truly independent system that does not rely on such external information. We present a method for the localization of such suitable (see [57]) illuminators, utilizing cooperative targets that transmit their ground truth information. Such a procedure was first described by Yi *et al.* [83].

In [79], Malanowski *et al.* show one possible implementation of such a technique. Here, we present alternative approaches to do so and examine their performance with numerical simulations, especially with respect to the standard deviation of the measurement error, which Malanowski identified as the main issue in the loss of performance when comparing real-world results to the simulation results of Yi *et al.* [83].

In real-world scenarios, we usually have to deal with multi-target situations. Therefore, a correct solution (or minimization) of the equations for the illuminator position requires that the observed range and Doppler information (from the radar measurements) are correctly assigned to the respective cooperative targets (with their known positions and velocities). Observations may even be discarded if the correct target cannot be assigned or the target ground truth is not known. Due to the complicated dependencies of bi-static range and Doppler from the 3-dimensional positions of the receiver, target, and (unknown) illuminator, and from the target velocity vector, this assignment has no obvious or direct solution. With the planar approximation method, we show a novel simplified approach for the association of the bistatic measurement results to the ground truth information, which is computationally efficient and which would be suitable for real-time implementation.

### 4.3.2. Ground Truth Acquisition

For our work, we use ADS-B (Automatic Dependent Surveillance-Broadcast) to acquire the ground truth information of the cooperative targets [84]. Since ADS-B transmitters are installed on most commercial aircraft, we found a multitude of cooperative targets at all our test sites so far. The ADS-B signals are received with an additional RTL2832U based SDR independently from the passive radar system, and all incoming messages are logged while passive radar measurements are taken. It has to be noted that the ADS-B messages are not synchronized to the acquired passive radar measurements, therefore the ground truth information cannot be mapped to the range-Doppler targets directly. For the mapping, the position and speed vectors of the targets are interpolated with the two nearest available ADS-B positions[9]. We use a simple linear interpolation, which assumes that the aircraft is not maneuvering between two ADS-B updates. Of course, the unknown error resulting from this assumption adds further to the range-Doppler measurement error.

### 4.3.3. Implementation of the Illuminator Finder

We assume a dataset for each illuminator containing range-Doppler measurements with the corresponding ground truth from multiple cooperative targets, with several data points each. A procedure for assigning the range-Doppler measurements to the ground truth is

[9]ADS-B position data updates were irregular, varying from one update per second to approximately one update every 10 s.

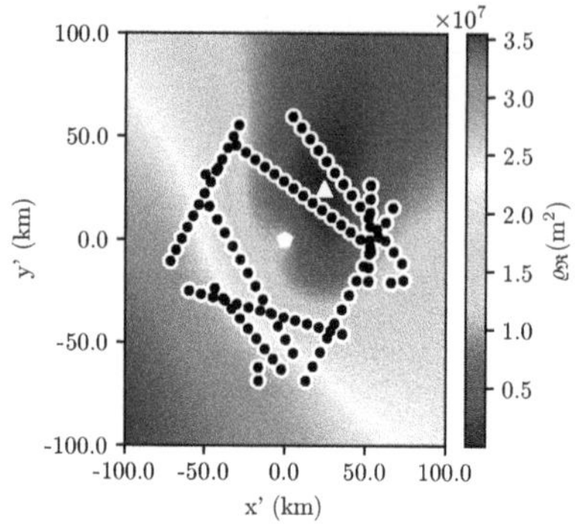


(a) Example of the cost function for range measurements.

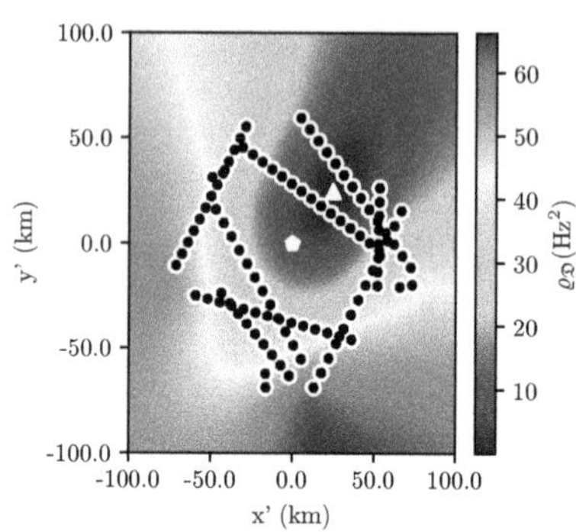


(b) Example of the cost function for Doppler measurements.

Figure 4.8.: Example of the cost functions. The black dots mark the simulated measurements; the white pentagon in the center marks the position of the receiver, and the white triangle marks the illuminator position.

described later. By applying a dataset, with a total of $N$ measurements, to Equation (2.17) or Equation (2.18), we can describe the cost functions for range and Doppler

$$\varrho_{\mathfrak{R}} = \frac{1}{N} \sum_{n=1}^{N} (\mathfrak{R}_n - \mathfrak{R}_{n,meas})^2 \tag{4.15}$$

$$\varrho_{\mathfrak{D}} = \frac{1}{N} \sum_{n=1}^{N} (\mathfrak{D}_n - \mathfrak{D}_{n,meas})^2. \tag{4.16}$$

Figures 4.8(a) and 4.8(b) show a plot of these functions for an area of 200 km by 200 km around the receiver. For both plots, the same set of randomly generated flights (black dots) are used. It is obvious that both the Doppler data and the range data[10] have a clear minimum at the position of the illuminator. Solving these equations for the illuminator position leads to an overdetermined non-linear system of equations, for which we will present three different solution approaches in further detail.

### Gauss-Newton Approximation

One possibility for solving the equations for the illuminator position is the Gauß-Newton iteration (GNI) algorithm [85, p. 254 ff.], which is an iterative procedure for solving non-linear least-squares (LS) problems [85, p. 245 ff.]. The general iteration of the algorithm is described by

$$x_{k+1} = x_k - \gamma \cdot \boldsymbol{J}^+ \cdot r, \tag{4.17}$$

where $r$ describes the residuum between the measured value of either range or Doppler and their calculated values. $\boldsymbol{J}$ is the corresponding Jacobian matrix[11], and $\gamma$ is a step width

[10] Or their combination as shown in [79].

[11] $\boldsymbol{J}^+$ describes the pseudoinverse $((\boldsymbol{J}^\mathrm{T}\boldsymbol{J})^{-1}\boldsymbol{J}^\mathrm{T})$.

$(0 < \gamma < 1)$ to ensure the convergence of the algorithm. This type of iterative solution furthermore needs an initial guess of the solution as a starting point. We show two implementations of this method, one for range values and one for Doppler. The first step in both is to move the receiver coordinates to the origin of the coordinate system by subtracting $\vec{B}$ from $\vec{T}$[12]. We get

$$\begin{aligned}\mathfrak{R} = &\sqrt{O_{\mathrm{x}}^2 + O_{\mathrm{y}}^2 + O_{\mathrm{z}}^2} - \sqrt{I_{\mathrm{x}}^2 + I_{\mathrm{y}}^2 + I_{\mathrm{z}}^2} \\ &+\sqrt{(I_{\mathrm{x}} - O_{\mathrm{x}})^2 + (I_{\mathrm{y}} - O_{\mathrm{y}})^2 + (I_{\mathrm{z}} - O_{\mathrm{z}})^2},\end{aligned} \tag{4.18}$$

with $\vec{I} = (I_{\mathrm{x}}, I_{\mathrm{y}}, I_{\mathrm{z}})^{\mathrm{T}}$ and $\vec{O} = \vec{T} - \vec{B}$, to describe the range of a target. The iteration is described by

$$\vec{I}_{k+1} = \vec{I}_k - \gamma \cdot \boldsymbol{J}_{\mathfrak{R}}^{+} \cdot \vec{r}_{\mathfrak{R}}. \tag{4.19}$$

The residuum is the difference between the calculated range value and the measured range value, which leads to the residuum vector

$$\vec{r}_{\mathfrak{R}} = \begin{bmatrix} \mathfrak{R}_1 - \mathfrak{R}_{1,\mathrm{meas}} \\ \vdots \\ \mathfrak{R}_N - \mathfrak{R}_{N,\mathrm{meas}} \end{bmatrix}.$$

**J** is given by

$$\boldsymbol{J} = \begin{bmatrix} \frac{\partial \mathfrak{R}_1}{\partial I_{\mathrm{x}}} & \frac{\partial \mathfrak{R}_1}{\partial I_{\mathrm{y}}} & \frac{\partial \mathfrak{R}_1}{\partial I_{\mathrm{z}}} \\ \vdots & \vdots & \vdots \\ \frac{\partial \mathfrak{R}_N}{\partial I_{\mathrm{x}}} & \frac{\partial \mathfrak{R}_N}{\partial I_{\mathrm{y}}} & \frac{\partial \mathfrak{R}_N}{\partial I_{\mathrm{z}}} \end{bmatrix}.$$

The equations for the Doppler method can be developed in a similar way, starting from

$$\begin{aligned}\mathfrak{D} = \frac{f}{c} \cdot \Bigg( & v_{\mathrm{x}} \cdot \frac{-T_{\mathrm{x}}}{\sqrt{T_{\mathrm{x}}^2 + T_{\mathrm{y}}^2 + T_{\mathrm{z}}^2}} \\ & + v_{\mathrm{y}} \cdot \frac{-T_{\mathrm{y}}}{\sqrt{T_{\mathrm{x}}^2 + T_{\mathrm{y}}^2 + T_{\mathrm{z}}^2}} + v_{\mathrm{z}} \cdot \frac{-T_{\mathrm{z}}}{\sqrt{T_{\mathrm{x}}^2 + T_{\mathrm{y}}^2 + T_{\mathrm{z}}^2}} \\ & - v_{\mathrm{x}} \cdot \frac{T_{\mathrm{x}} - I_{\mathrm{x}}}{\sqrt{(T_{\mathrm{x}} - I_{\mathrm{x}})^2 + (T_{\mathrm{y}} - I_{\mathrm{y}})^2 + (T_{\mathrm{z}} - I_{\mathrm{z}})^2}} \\ & - v_{\mathrm{y}} \cdot \frac{T_{\mathrm{y}} - I_{\mathrm{y}}}{\sqrt{(T_{\mathrm{x}} - I_{\mathrm{x}})^2 + (T_{\mathrm{y}} - I_{\mathrm{y}})^2 + (T_{\mathrm{z}} - I_{\mathrm{z}})^2}} \\ & - v_{\mathrm{z}} \cdot \frac{T_{\mathrm{z}} - I_{\mathrm{z}}}{\sqrt{(T_{\mathrm{x}} - I_{\mathrm{x}})^2 + (T_{\mathrm{y}} - I_{\mathrm{y}})^2 + (T_{\mathrm{z}} - I_{\mathrm{z}})^2}} \Bigg).\end{aligned} \tag{4.20}$$

To evaluate the reliability of this approach and to find reasonable values for $\gamma$, this approach is tested with simulated data[13]. Furthermore, the influence of the number of targets on

[12] Where $\vec{B}$ describes the position of the receiver, $\vec{T}$ the position of the target and $\vec{I}$ the position of the illuminator. See also Figure 2.1.

[13] For the simulation, the receiver and the illuminator are defined at fixed positions, 10 km apart. The position of the targets and the initial guess for the illuminator position are chosen at random for each simulation. The range and Doppler values are calculated with Equation (2.17) and Equation (2.18). No additional measurement error is added to the calculated values. The target position $\vec{T} = (T_{\mathrm{x}}, T_{\mathrm{y}}, T_{\mathrm{z}})^{\mathrm{T}}$ is chosen with random cartesian coordinates around the receiver position $\vec{B} = (0, 0, 0)^{\mathrm{T}}$, with $T_{\mathrm{x}}$, $T_{\mathrm{y}}$ and $T_{\mathrm{z}}$ each between -10 km and 10 km.

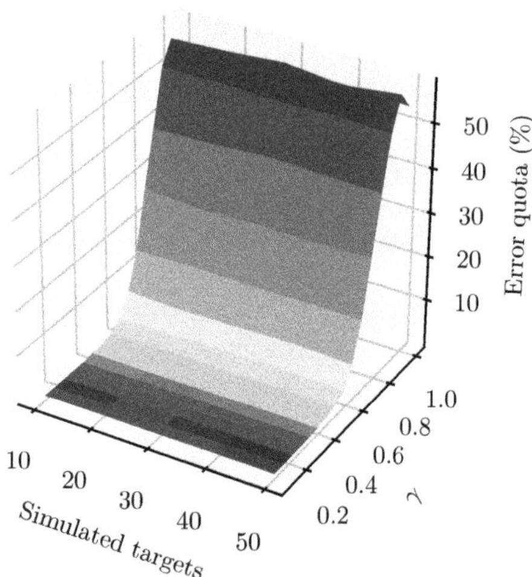


Figure 4.9.: Plot of the influence of the number of simulated targets and the step-width $\gamma$ on the error quota of the range based GNI method. Note, that the error quota depicted in this plot only converges to zero, but never reaches it.

the reliability is examined. For each combination of $\gamma$ and measurement count, 10 000 simulations are evaluated. A simulation run is counted as successful if the GNI reaches the illuminator position with a distance smaller than 1 m. Otherwise, the simulation is counted as failed. Figure 4.9 shows the percentage of failed simulations. It can be seen that the error quota does not depend on the number of targets used in the simulation and that a $\gamma$ value below 0.5 yields the best results. The data furthermore shows that even with ideal, simulated data without added error, the algorithm fails in a small percentage of cases to find the correct illuminator position. This behavior can be described by local minima in non-linear optimization.

#### Spherical-Interpolation Method

Another method to determine the illuminator position by using Equation (4.18) is shown in [86] and is called the Spherical-Interpolation (SI) Method. The difference between Malanowski's algorithm in [86] and our approach is that the target and the illuminator have interchanged their roles. For a single measurement $n$ the target position is described by the vector $\vec{T_n} = (T_{x,n}, T_{y,n}, T_{z,n})^{\mathrm{T}}$. The corresponding range is denoted as $\mathfrak{R}_n$. Before the SI method can be applied, Equation (4.18) has to be rearranged to

$$\begin{aligned} 2(O_{x,n} I_{\mathrm{x}} + O_{y,n} I_{\mathrm{y}} + O_{z,n} I_{\mathrm{z}}) &= \\ O_{x,n}^2 + O_{y,n}^2 + O_{z,n}^2 - w_n^2 + 2 w_n P, \end{aligned} \tag{4.21}$$

with $w_n = \Re_n - \sqrt{O_{x,n}^2 + O_{y,n}^2 + O_{z,n}^2}$, the unknown position of the illuminator $\vec{I} = (I_\mathrm{x}, I_\mathrm{y}, I_\mathrm{z})^\mathrm{T}$ and

$$P = \|\vec{I}\|. \tag{4.22}$$

For a dataset with a total number of $N$ measurements, Equation (4.21) can be written in matrix vector notation as

$$2\boldsymbol{S}\vec{I} = \vec{V} + 2P\vec{W}, \tag{4.23}$$

with $\vec{W} = \left[w_1 \cdots w_N\right]^\mathrm{T}$,

$$\boldsymbol{S} = \begin{bmatrix} O_{x,1} & O_{y,1} & O_{z,1} \\ \vdots & \vdots & \vdots \\ O_{x,N} & O_{y,N} & O_{z,N} \end{bmatrix},$$

$$\vec{V} = \begin{bmatrix} O_{x,1}^2 + O_{y,1}^2 + O_{z,1}^2 - w_1^2 \\ \vdots \\ O_{x,N}^2 + O_{y,N}^2 + O_{z,N}^2 - w_N^2 \end{bmatrix}.$$

For $N > 3$, Equation (4.23) describes an over determined system of equations. We now assume that the value of $P$ is known and define an error vector

$$\vec{\epsilon}_\mathrm{I} = \vec{V} + 2P\vec{W} - 2\boldsymbol{S}\vec{I}, \tag{4.24}$$

which has to be minimized with respect to $\vec{I}$ in terms of LS. The solution to this minimization problem is not exact but unambiguous and yields [86]

$$\tilde{\vec{I}} = \frac{1}{2}\boldsymbol{S}^{+}(\vec{V} + 2P\vec{W}), \tag{4.25}$$

with the pseudoinverse matrix $\boldsymbol{S}^{+}$. To get a solution for $P$, we substitute Equation (4.25) into Equation (4.24), which results in a new error vector [86]

$$\vec{\epsilon}_P = (\boldsymbol{I} - \boldsymbol{S}\boldsymbol{S}^{+})(\vec{V} + 2P\vec{W}),$$

which now has to be minimized with respect to $P$. The solution for $\tilde{P}$[14] is given by [86]

$$\tilde{P} = -\frac{1}{2}\frac{\vec{W}^\mathrm{T}\boldsymbol{T}\vec{V}}{\vec{W}^\mathrm{T}\boldsymbol{T}\vec{W}}, \tag{4.26}$$

with $\boldsymbol{T} = \boldsymbol{I} - \boldsymbol{S}\boldsymbol{S}^{+}$. Now, Equation (4.18) can be solved by calculating $\tilde{P}$ with Equation (4.26) and using this result in Equation (4.25) to get the illuminator position $\tilde{\vec{I}}$. It is clear that, compared to the GNI approximation (see 4.3.3), this represents a closed solution to the problem, so no initial guess is necessary for this procedure.

[14] The tilde again denotes that this solution is a solution in terms of LS.

### Spherical-Intersection Method

In [86], Malanowski describes the Spherical-Intersection Method (SX), another variant of a closed-form solution. Starting from Equation (4.25) we substitute

$$\boldsymbol{a} = \frac{1}{2}\boldsymbol{S}^{+}\vec{V}, \quad \boldsymbol{b} = \boldsymbol{S}^{+}\vec{W}$$

to get

$$\tilde{\vec{I}} = \boldsymbol{a} + \boldsymbol{b}P. \tag{4.27}$$

If we now substitute $\vec{I}^{\mathrm{T}}\vec{I} = ||\vec{I}||^2$ and insert this into Equation (4.22), we get the quadratic equation

$$(\boldsymbol{b}^{\mathrm{T}}\boldsymbol{b} - 1)P^2 + 2\boldsymbol{a}^{\mathrm{T}}\boldsymbol{b}P + \boldsymbol{a}^{\mathrm{T}}\boldsymbol{a} = 0. \tag{4.28}$$

The two solutions of Equation (4.28) are given by

$$P_{1,2} = \frac{-\boldsymbol{a}^{\mathrm{T}}\boldsymbol{b} \pm \sqrt{(\boldsymbol{a}^{\mathrm{T}}\boldsymbol{b})^2 - (\boldsymbol{b}^{\mathrm{T}}\boldsymbol{b} - 1)\boldsymbol{a}^{\mathrm{T}}\boldsymbol{a}}}{\boldsymbol{b}^{\mathrm{T}}\boldsymbol{b} - 1}. \tag{4.29}$$

With Equation (4.29) and Equation (4.25), we can calculate two possible locations for the illuminator. The correct solution can be chosen by comparison of the value of the cost function Equation (4.15); the position with the lowest cost is used [86]. If due to larger measurement errors, the discriminant of Equation (4.28) is less than zero, the solution of Equation (4.29) has an imaginary component. In this case, only the real component of the solution is used [86].

### Initial Data Set Matching Using Planar Approximation

The approaches described above require a dataset with range-Doppler measurements associated with a ground truth coordinate. A possible solution to solve this target association ambiguity is proposed by Malanowski in [79]; the proposed method checks for each combination of bistatic measurements with available ground truth information the minimum value of the cost function. However, for larger data sets, this procedure becomes extremely computationally complex and is therefore no longer applicable. To solve this problem, we are suggesting a new approach.

In our system, the matching of a range-Doppler target to the ground truth coordinates is done by comparison of the true bearing, calculated from the target coordinate, and the bearing extracted from the beamforming process. To match a combination, we allow for a deviation of $\pm 2.5°$. This simple approach manages to filter out a lot of false combinations, but some unwanted, incorrect assignments are still left. These false associations lead to problems with the calculation of the illuminator position, which are described in Chapter 4.3.5.

What we need at this point is a good approximation of the illuminator position to filter out the incorrectly assigned measurement results from our data set. To this end, we conceived the planar approximation (PA) method as a robust procedure simplifying the problem to a single-variable problem by further utilizing the measured bearings toward the illuminators[15]. For that purpose, we assume that the illuminator is on a tangent plane to the earth originating at the receiver location, thereby ignoring its actual height above ground and the curvature of the earth. We can describe this plane with a coordinate system, whose z-axis is the extension of the vector $\vec{B}$, as can be seen in Figure 4.10. The y-axis of this system points towards the North Pole. In this coordinate system, the position of the illuminator $\vec{I'}$ can be described with the measured bearing $\alpha$ and its distance $R$:

$$\vec{I'} = \begin{bmatrix} R \cdot \cos\alpha \\ R \cdot \sin\alpha \\ 0 \end{bmatrix}. \tag{4.30}$$

With this simple approach, we can calculate an illuminator position for every single measurement we have, without the necessity for combining multiple measurements, which may bear the risk of adding false assignments to our dataset. Incorrectly assigned range-Doppler measurements would lead to outliers of the calculated position on the straight line given by the angle towards the illuminator. The correctly assigned measurement results appear distributed around the actual location of the illuminator. To minimize the outliers' influence, the illuminator position can now be formed from this distribution with the median instead of simply calculating the mean value. Since the position of the targets and the receiver location are described in a global earth-centered, earth-fixed (ECEF) coordinate system, we decided to transform this local coordinate system to the global one. The transformation is described by

$$\vec{I} = \vec{B} + I'_{\mathrm{x}} \cdot \vec{e}'_{\mathrm{x}} + I'_{\mathrm{y}} \cdot \vec{e}'_{\mathrm{y}} + I'_{\mathrm{z}} \cdot \vec{e}'_{\mathrm{z}}, \tag{4.31}$$

with the unit vectors

$$\vec{e}'_{\mathrm{x}} = \begin{bmatrix} -\sin\varphi \\ \cos\varphi \\ 0 \end{bmatrix}, \quad \vec{e}'_{\mathrm{y}} = \begin{bmatrix} -\cos\theta\cos\varphi \\ -\cos\theta\sin\varphi \\ \sin\theta \end{bmatrix}, \quad \vec{e}'_{\mathrm{z}} = \frac{\vec{B}}{\left\|\vec{B}\right\|}. \tag{4.32}$$

The angles $\theta$ and $\varphi$ are described by

$$\begin{aligned} \theta &= \arccos\left(\frac{B_{\mathrm{z}}}{\sqrt{B_{\mathrm{x}}^2 + B_{\mathrm{y}}^2 + B_{\mathrm{z}}^2}}\right), \\ \varphi &= \operatorname{arctan2}(B_{\mathrm{y}}, B_{\mathrm{x}}), \end{aligned} \tag{4.33}$$

with arctan2 being defined in the usual way[16]. Applying Equation (4.32) to Equation (4.31) results in

$$\vec{I} = \begin{bmatrix} B_{\mathrm{x}} - R\cos\alpha\sin\varphi - R\sin\alpha\cos\theta\cos\varphi \\ B_{\mathrm{y}} + R\cos\alpha\cos\varphi - R\sin\alpha\cos\theta\sin\varphi \\ B_{\mathrm{z}} + \sin\alpha\sin\theta \end{bmatrix}. \tag{4.34}$$

By applying Equation (4.34) to Equation (4.18), we get an equation for the range with only one unknown variable, $R$. This equation $\mathfrak{R}(R)$ can be solved iteratively using Equation

[15] The bearing toward the illuminator is determined with the same beamformer and sinc-interpolation process as described for the targets. It is assumed that the reference signal is the dominant component in the directly received signal, which is the case in scenarios with line-of-sight towards the transmitter. See 4.2.3.

[16] See e.g. the arctan2 documentation on numpy.org or other numerical tools, the C language ISO/IEC standard 9899, or floating point standards such as IEEE-754 or its counterpart ISO/IEC 60559.

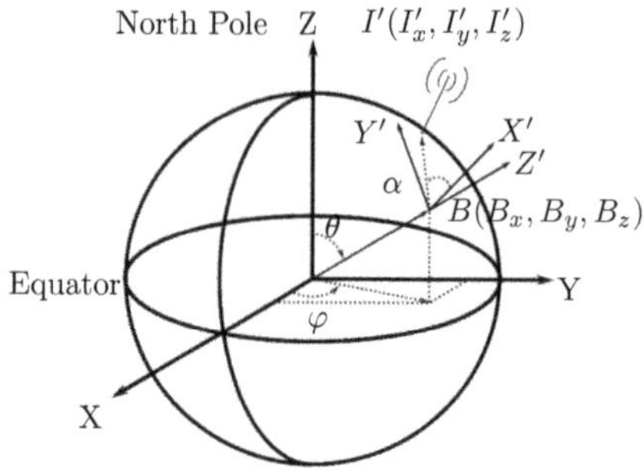


Figure 4.10.: Overview of the used coordinate systems. The ECEF coordinate system is marked by $X, Y, Z$. The local system, tangential to the surface of the earth is marked as $X', Y', Z'$.

Table 4.1.: Influence of the PA method on the GNI based methods

| Value | Range | | Doppler | |
|---|---|---|---|---|
| | without PA | with PA | without PA | with PA |
| Mean error | 5.604 m | 1.606 µm | 65.104 m | 1.652 µm |
| Standard dev. | 223.9384 m | 1.102 µm | 902.476 m | 1.252 µm |
| Max error | 12.343 km | 10.914 µm | 20.359 km | 27.741 µm |
| Error quota | 0.10 % | 0 % | 0.57 % | 0 % |

(4.17) to calculate a value $R$ for each measurement, resulting in an illuminator position using Equation (4.34). By taking the median of all these calculated positions, quite a good guess of the illuminator's position can be made. Figure 4.11 shows an example of this process, using a normally distributed range error with a standard deviation of $\sigma = 300\,\text{m}$. First, this guess of the illuminator's position is used to filter the dataset further and remove false combinations of range-Doppler and ground truth measurements. This is done using Equation (4.18) and Equation (4.20) to calculate the range and Doppler values, using the estimated illuminator position and the ground truth data. The results are compared to the measured range and Doppler values from the radar. If they differ too much, the combination is removed. Second, the illuminator position is used as the initial guess for the GNI approximation, which now uses the double-filtered dataset. As a result of this more precisely known starting position, the number of converging solutions increases.

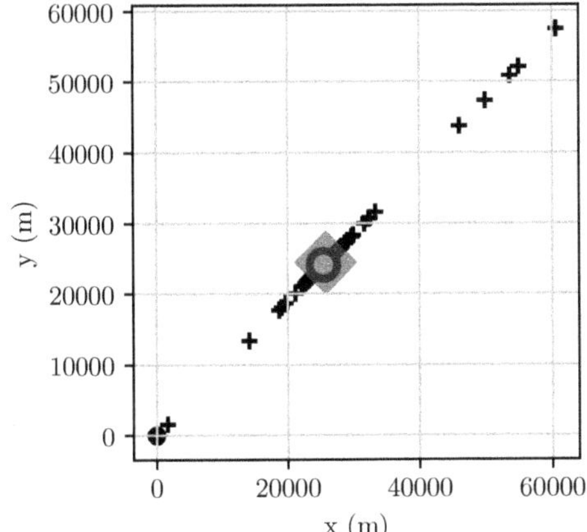


Figure 4.11.: Example of the planar approximation (PA). The black dot marks the receiver position; the black circle marks the actual illuminator position. The plus markers show the calculated position for each individual measurement, and the grey diamond marks the median of these positions.

### 4.3.4. Evaluation of the Methods with Simulated Data

In this section, we will evaluate the different presented methods based on randomly simulated scenarios[17]. First, we examine the effect of using the position calculated with the PA as an initial guess for the GNI based methods. For that purpose, 10 000 simulations for each of the four combinations, GNI either based on range or Doppler and using the initial guess from the PA or a random position, are conducted. To take measurement error in the angular estimation into account, a uniformly distributed random error of $\pm 2°$ is added to the bearing $\alpha$. No additional error is added to the range-Doppler data for this examination. Table 4.1 shows the results.[18] The results shown make it clear that, although the error probability is low, the standalone GNI methods are unreliable as the errors can exceed 10 km. These errors are mitigated completely by combining the GNI method with PA, thus severely improving their reliability.

In the next step, we examine the priorly neglected influence of the range/Doppler meas-

[17]For all simulations, the illuminator position is assumed at a fixed position, 35 km away from the receiver with the antenna position 100 m above the ground. The angle $\alpha$ describes the bearing from the receiver towards the illuminator. The targets are simulated as commercial airliners, with linear flight path intersecting with the measurement area. To simulate these flights, the start and end positions are chosen at random, both in a range between 50 km and 80 km around the receiver, with an altitude between 2 km and 12 km. The flights are modeled as a linear movement with constant velocity from the start point to the endpoint. The velocity is chosen randomly between 214 m/s and 286 m/s, resembling typical airliner cruising speeds. To get viable distances between the measurements, a time step of 30 s is used. For each simulated data set, flights are generated until a set of 100 measurements is complete. A single measurement consists of the target position, its speed, and the range-Doppler value.

[18]The error already described in 4.3.3 appears in the error quota, which is again defined as the percentage of calculations with an error bigger than 1 m.

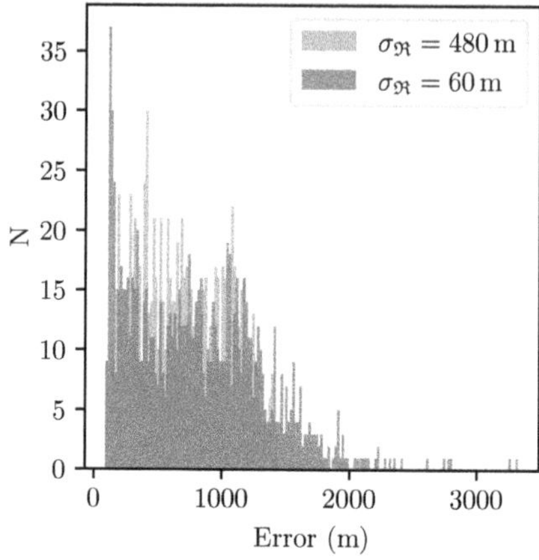


Figure 4.12.: Error Distribution for the PA method with different error distributions for the measured range and an random angle error of $\pm 2°$.

urement error. Therefore, a normal-distributed error is added to the calculated range and Doppler values from the simulated radar. The standard deviations $\sigma_{\mathfrak{R}}$ and $\sigma_{\mathfrak{D}}$ of this error are varied in this investigation, and for each error distribution, 1000 simulations are conducted. The added error of the angular measurement is still $\pm 2°$. The distance from the calculated position to the actual illuminator position is evaluated as a measure of the method's performance. For better illustration, the results are shown both in histograms with two exemplary results in Figures 4.12 - 4.13(d) and as a graph with the mean error and the standard deviation of the results in Figure 4.14.

To make the results comparable, each set of simulated targets is tested with all described methods. The first method applied to the data set is the PA. Figure 4.12 and 4.14 show that the PA method is fairly resistant against added range error. For range errors with an error distribution of $\sigma_{\mathfrak{R}} > 500\,\mathrm{m}$, the mean error is approximately constant. For smaller values, it is even slightly larger, which could be a result of the non-linear median filtering. The PA results are then used as an initial guess for the GNI methods; the outcome of these investigations can be found in Figures 4.13(a) and 4.13(b). In both cases, the mean error for the calculated illuminator position grows linearly with the added measurement error (see Figure 4.14). A remarkable difference between the two methods can be found in the progression of the standard deviation compared to the mean value. For the range-based methods, the standard deviation is always lower than the mean error, but for the Doppler-based GNI method, it is partly larger and overall closer to the value of the mean error. This means that for comparable mean errors, the results of the Doppler based method has a greater spread, making the range-based methods more reliable. The results for the SI and SX methods are shown in Figure 4.13(c) and Figure 4.13(d). The mean error increases linearly with the error on the input data (see Figure 4.14). However, in both cases, the slope is considerably larger than with the GNI method. For the SI method, it is nearly doubled, for the SX method nearly tripled. The behavior of the deviation is comparable to the GNI method.

Table 4.2.: Results without PA-filtering

| Method | Position error | unit |
|---|---|---|
| PA | 1.361 | km |
| GNI (range) | $> 1e18$ | km |
| GNI (Doppler) | $> 2e20$ | km |
| SI (range) | 57.409 | km |
| SX (range) | 30.385 | km |

The advantages of the GNI based methods can be explained by approximations in the closed solutions of the SI and SX methods [87, 88]. Finally, the influence of the angular error, previously modeled at $\pm 2°$, on the PA and the dependent GNI procedures is investigated. For this purpose, simulations are performed with a range error of $\sigma_{\mathfrak{R}} = 480\,\text{m}$ and a Doppler error of $\sigma_{\mathfrak{D}} = 2.8\,\text{Hz}$, respectively, and a fixed angular error is added to the angle towards the illuminator. For each variation of the angular error, 1000 simulations are conducted. In Figure 4.15, it can be seen that with increasing angular error the error of the PA method also increases linearly. However, the performance of the GNI methods is not affected since the result of the PA is merely used as an initial value of those methods.

Summing up these results, the GNI, in combination with the PA method, yields the best results for all considered measurement error distributions.

### 4.3.5. Real-world Example

The methods described above are now tested with real-world measurements, which were carried out in the Netherlands near The Hague. An illuminator[19] transmitting FM Audio broadcast at 95.2 MHz is chosen.[20] As described above, commercial flights transmitting ADS-B are employed as cooperative targets.

The dataset is generated from the range-Doppler data by applying a CFAR with a 7.5 dB SNR threshold. The targets are then matched to interpolated ADS-B targets by the measured angle, allowing for a maximum deviation of $\pm 2.5°$[21]. In this case, an illuminator whose position is known is chosen to validate the results: for the sake of the experiment it is assumed to be unknown. First, we apply the above-described methods directly to this dataset, which yields the results described in Table 4.2. It is apparent that all methods, except the PA, fail to get near the illuminator position due to the above-described incorrectly assigned targets. Looking at the presentation of the cost functions (4.15) and (4.16) shown in Figure 4.16(a) and 4.16(b), it becomes clear why these methods fail. For neither of the two data sets, the cost function has a minimum at the actual position of the illuminator; on the contrary, the Doppler function seems to have it's maximum close to the position of the illuminator[22].

[19] The respective transmitter is located at 52°08′13.6″ N 4°38′47.3″ E, approximately 90 m above ground. It is part of the "Zendmast Bezuidenhout" broadcast tower [89].

[20] FM broadcast services operates with a bandwidth $B$ of about 100 kHz, resulting in maximum range resolution $\Delta R = \frac{c_0}{2B}$ of approximately 3 km.

[21] The threshold value for the angle is determined empirically based on the available measurement data and probably overestimates the system's actual measurement accuracy. It is deliberately chosen small in order to strongly preselect the data set for the following processing .

[22] This also explains why the two GNI methods do not reach a minimum and apparently only increase the distance in each iteration, leading to the astronomical error sizes (1e18 km corresponds approximately to the diameter of the Milky Way).

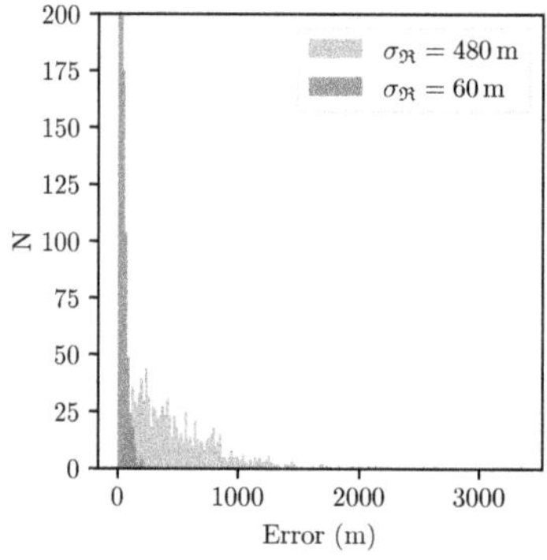


(a) Error distribution for the range-based GNI with different error distributions for the measured range, with the initial guess calculated by the PA method.

(b) Error distribution for the Doppler-based GNI with different error distributions for the measured Doppler, with the initial guess calculated by the PA method.

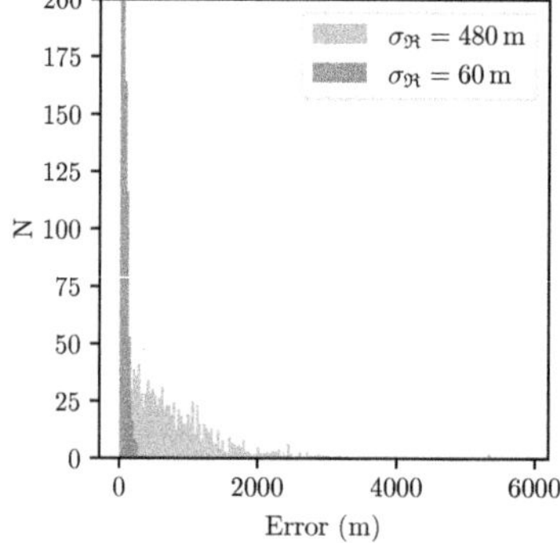


(c) Error distribution for the range-based SI method with different error distributions for the measured range.

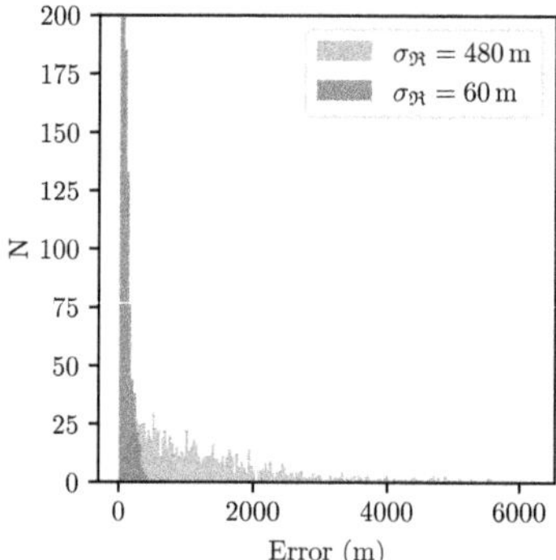


(d) Error distribution for the range-based SX method with different error distributions for the measured range.

Figure 4.13.: Distribution of simulation results as histograms.

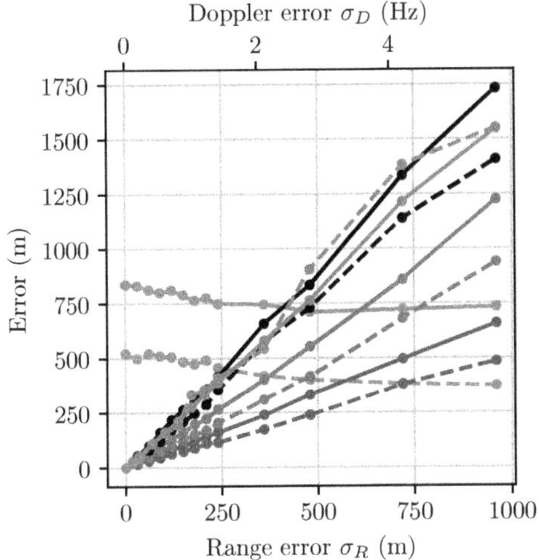


Figure 4.14.: Deviation of the calculated position for the PA (red), the GNI with range (blue) / Doppler (green) data, the SI (magenta), and the SX (black). The solid lines represent the mean error, the dotted lines the standard deviation.

To filter these false associations from the data set, we use the result of the PA method. Based on these results, we calculate range-Doppler values for each target. These values are then compared with the assigned measurement data, and any deviating values are removed from the data set.[23]

The results achieved with these filtered datasets are shown in Table 4.3. The position found with the PA is again used as an initial guess for both GNI methods.

Table 4.3.: Results with PA-filtering

| Method | Position error | unit |
|---|---|---|
| PA | 1.361 | km |
| GNI (range) | 0.67 | km |
| GNI (Doppler) | 3.252 | km |
| SI (range) | 0.456 | km |
| SX (range) | 0.782 | km |

The range-data-based methods improve drastically when based on the filtered data set, while the improvement of the Doppler-data-based method is not as impressive, as the error is comparatively large. All range-data-based methods produce an error smaller than 1 km. The best result for the real-world data is achieved by Malanowski's SI method [86], reaching

[23] For this measurement we are working with a threshold of 500 m range deviation and 5 Hz Doppler deviation.

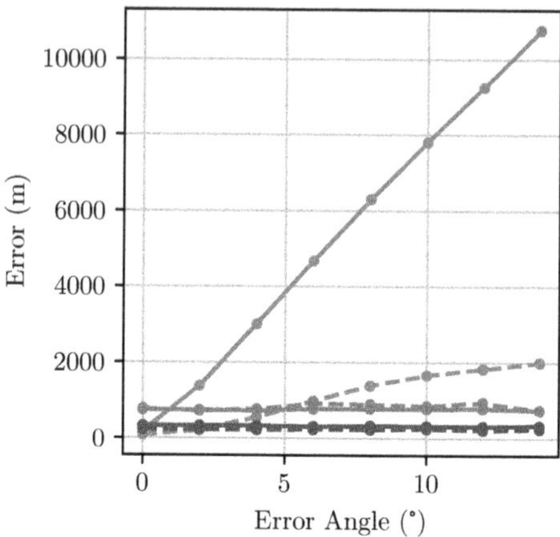


Figure 4.15.: Deviation of the calculated position plotted over varying errors in the bearing towards the illuminator. Results are shown for the PA (red) and the GNI with range (blue) / Doppler (green) data. The solid lines represent the mean error, the dotted lines the standard deviation.

the illuminator position with an error of less than 500 m. To illustrate the improvement of the data set, we look again at the representation of the cost function in Figure 4.16(c) and 4.16(d). For the range data in Figure 4.16(c), it is now clear that the minimum is at the illuminator position. For the Doppler data in Figure 4.16(d), the position of the minimum is not so clear, but at least the illuminator position is now in a range of minimum values, even if these extent over a large range. We assume that the measurement error in the Doppler data is much greater compared to the range data. Since the error in the data is a combined error between the actual measurement error and the error within the interpolated ADS-B data, it is not possible to analyze the origin of this error in further detail.

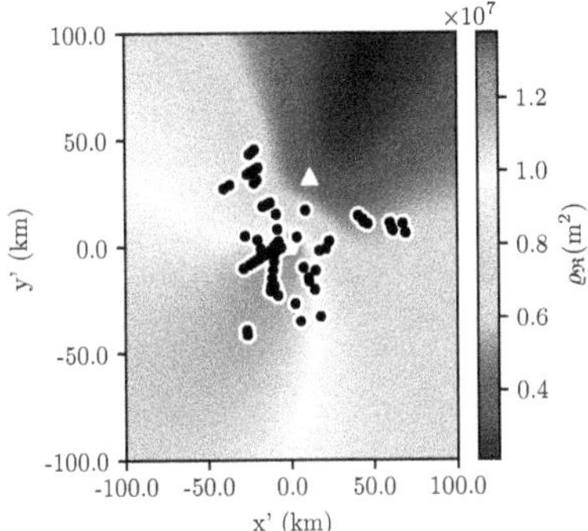


(a) The plot of the cost function for the real-world range measurements.

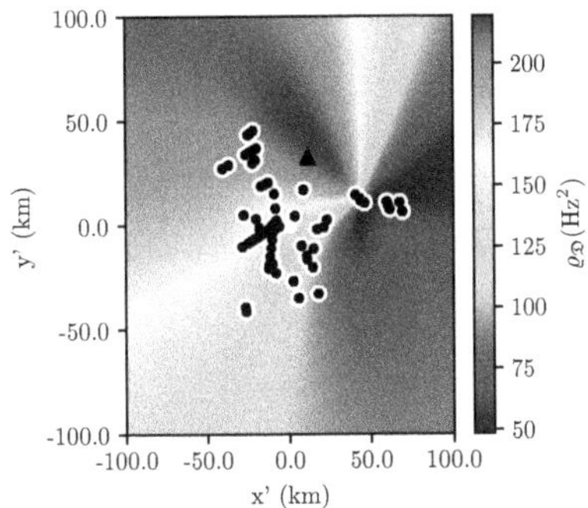


(b) The plot of the cost function for the real-world Doppler measurements.

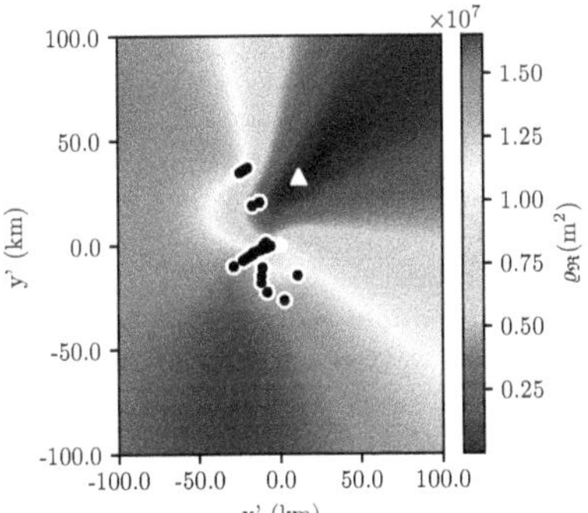


(c) The plot of the cost function for the real-world range measurements, after filtering based on the PA results.

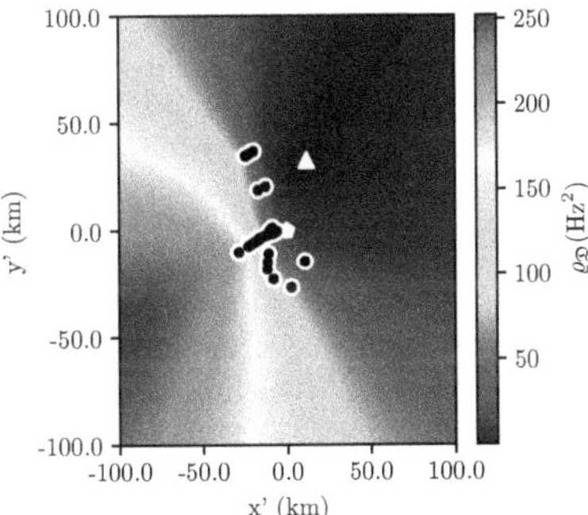


(d) Plot of the cost function for the real-world Doppler measurements, after filtering based on the PA results.

Figure 4.16.: The cost functions based on the real measurements with and without prior PA-filtering of the data. The black dots mark the used measurements; the white pentagon in the center marks the position of the receiver, and the triangle marks the illuminator position.

### 4.3.6. Conclusion

We have presented different methods to find the position of an unknown illuminator of opportunity in a passive radar scenario using the ADS-B information of cooperative targets and evaluated their performance with simulated data sets. All of these approaches fail in a real-world scenario due to false-positive combinations of radar observations with ground truth data. To solve this problem, we devised the PA method, which yields a robust approximation of the illuminator position. We have demonstrated that this approximation can be used to pre-filter a dataset containing falsely matched targets and, furthermore, can be used as an initial guess for iteration-based solving approaches, improving their reliability. We have shown that our technique improves the performance of iteration-based solving methods with simulated data sets and that it is a robust approach for solving the target association ambiguity problem in a real-world data set.

## 4.4. Target Tracking

The task of the passive radar system is to determine the position and speed of aircraft. This information can be calculated for a target from Equation (2.17) and Equation (2.18) if at least three range-Doppler measurements from independent illuminators are available for the target. These equations can be solved, for example, using the GNI procedure described in 4.3.3. However, in a practical implementation, further challenges arise.
On the one hand, the association problem emerges as soon as more than one target is present in the monitored area. On the other hand, it cannot be ensured that a target can always be detected with all illuminators. On the contrary, due to the distribution of the illuminator systems and the high dynamic range in the bistatic RCS, it can be assumed that targets can often only be measured with only one illuminator, if at all.
A tracking algorithm is used to overcome this issue. It creates a simple model for each potential target. With this model, predictions of the target's state are possible, enabling the processing of asynchronous or incomplete data sets. The problem of the initial mapping of measurements to targets needed to start such a track is solved with a modified version of the PA algorithm described in Chapter 4.3.3. The core of the tracking algorithm is a UKF (see 2.9.2), which is initialized for each new track. The algorithm itself is divided into the following five phases:

- Target estimation and clustering of measurements: For the measurements, target positions are estimated using the PA. Within these estimated positions, clusters of potentially related measurements are searched.
- Track prediction: The new position of the track is predicted from its previous state and if possible, the measurements are associated to existing tracks, or new track hypotheses are generated.
- Track update: The tracks are updated with the new measurements.
- Track initiation: New tracks are initiated from the measured data
- Track deletion: Old tracks that are no longer updated must be deleted to prevent "phantom tracks".

### 4.4.1. Tracking Filter Design

As mentioned above, each track is represented by a UKF. The current state of this filter is given by the state vector of position and velocity

$$\vec{x} = \begin{bmatrix} T_{\mathrm{x}} \\ v_{\mathrm{x}} \\ T_{\mathrm{y}} \\ v_{\mathrm{y}} \\ T_{\mathrm{z}} \\ v_{\mathrm{z}} \end{bmatrix}, \tag{4.35}$$

where $T_{x,y,z}$ describes the target's position and $v_{x,y,z}$ its velocity (see Figure 2.1) and the corresponding covariance matrix

$$\boldsymbol{P} = \begin{bmatrix} \sigma^2_{T_{\mathrm{x}}} & \sigma_{T_{\mathrm{x}},v_{\mathrm{x}}} & \sigma_{T_{\mathrm{x}},T_{\mathrm{y}}} & \sigma_{T_{\mathrm{x}},v_{\mathrm{y}}} & \sigma_{T_{\mathrm{x}},T_{\mathrm{z}}} & \sigma_{T_{\mathrm{x}},v_{\mathrm{z}}} \\ \sigma_{v_{\mathrm{x}},T_{\mathrm{x}}} & \sigma^2_{v_{\mathrm{x}}} & \sigma_{v_{\mathrm{x}},T_{\mathrm{y}}} & \sigma_{v_{\mathrm{x}},v_{\mathrm{y}}} & \sigma_{v_{\mathrm{x}},T_{\mathrm{z}}} & \sigma_{v_{\mathrm{x}},v_{\mathrm{z}}} \\ \sigma_{T_{\mathrm{y}},T_{\mathrm{x}}} & \sigma_{T_{\mathrm{y}},v_{\mathrm{x}}} & \sigma^2_{T_{\mathrm{y}}} & \sigma_{T_{\mathrm{y}},v_{\mathrm{y}}} & \sigma_{T_{\mathrm{y}},T_{\mathrm{z}}} & \sigma_{T_{\mathrm{y}},v_{\mathrm{z}}} \\ \sigma_{v_{\mathrm{y}},T_{\mathrm{x}}} & \sigma_{v_{\mathrm{y}},v_{\mathrm{x}}} & \sigma_{v_{\mathrm{y}},T_{\mathrm{y}}} & \sigma^2_{v_{\mathrm{y}}} & \sigma_{v_{\mathrm{y}},T_{\mathrm{z}}} & \sigma_{v_{\mathrm{y}},v_{\mathrm{z}}} \\ \sigma_{T_{\mathrm{z}},T_{\mathrm{x}}} & \sigma_{T_{\mathrm{z}},v_{\mathrm{x}}} & \sigma_{T_{\mathrm{z}},T_{\mathrm{y}}} & \sigma_{T_{\mathrm{z}},v_{\mathrm{y}}} & \sigma^2_{T_{\mathrm{z}}} & \sigma_{T_{\mathrm{z}},v_{\mathrm{z}}} \\ \sigma_{v_{\mathrm{z}},T_{\mathrm{x}}} & \sigma_{v_{\mathrm{z}},v_{\mathrm{x}}} & \sigma_{v_{\mathrm{z}},T_{\mathrm{y}}} & \sigma_{v_{\mathrm{z}},v_{\mathrm{y}}} & \sigma_{v_{\mathrm{z}},T_{\mathrm{z}}} & \sigma^2_{v_{\mathrm{z}}} \end{bmatrix}. \tag{4.36}$$

As described in 2.9.2, the UKF uses the SUT to generate a swarm of sigma-points $\mathcal{S} = \{w, \mathcal{X}\}$ from this state. For the SUT the parameters $\alpha = 0.1$, $\beta = 2$ and $\kappa = 0$ are used[24]. Using these parameters, the weights $w_i$ and the state sigma-points $\mathcal{X}$ are calculated according to the equations described in Chapter 2.9.2.
Next, the transition function is set up as

$$f(\mathcal{X}_{t-1}) = \begin{bmatrix} 1 & \delta t & 0 & 0 & 0 & 0 \\ 0 & 1 & 0 & 0 & 0 & 0 \\ 0 & 0 & 1 & \delta t & 0 & 0 \\ 0 & 0 & 0 & 1 & 0 & 0 \\ 0 & 0 & 0 & 0 & 1 & \delta t \\ 0 & 0 & 0 & 0 & 0 & 1 \end{bmatrix} \mathcal{X}_{t-1}, \tag{4.37}$$

where $\delta t$ is the time difference to the last calculated state vector. Following the practical example from Chapter 2.8, the error is described with three velocity components at the time of sampling. This yields the matrix of the process noise

$$\boldsymbol{Q} = \begin{bmatrix} \sigma^2_{v_x,\mathrm{err}}\delta t^2 & \sigma^2_{v_x,\mathrm{err}}\delta t & 0 & 0 & 0 & 0 \\ \sigma^2_{v_x,\mathrm{err}}\delta t & \sigma^2_{v_x,\mathrm{err}} & 0 & 0 & 0 & 0 \\ 0 & 0 & \sigma^2_{v_y,\mathrm{err}}\delta t^2 & \sigma^2_{v_y,\mathrm{err}}\delta t & 0 & 0 \\ 0 & 0 & \sigma^2_{v_y,\mathrm{err}}\delta t & \sigma^2_{v_y,\mathrm{err}} & 0 & 0 \\ 0 & 0 & 0 & 0 & \sigma^2_{v_z,\mathrm{err}}\delta t^2 & \sigma^2_{v_z,\mathrm{err}}\delta t \\ 0 & 0 & 0 & 0 & \sigma^2_{v_z,\mathrm{err}}\delta t & \sigma^2_{v_z,\mathrm{err}} \end{bmatrix}. \tag{4.38}$$

Where $\sigma^2_{v_{\mathrm{x}},\mathrm{err}}$, $\sigma^2_{v_{\mathrm{y}},\mathrm{err}}$, and $\sigma^2_{v_{\mathrm{z}},\mathrm{err}}$ each describe the variance of the expected velocity error. The spatial directions $x$, $y$ and $z$ refer to the ECEF coordinate system.
The last equation needed for setting up a UKF is the observation function that maps

[24] These values were chosen according to van der Meerwe's suggestions. See the quote in Section 2.9.2.

the state to the space of the measured parameters. It results from Equations (2.17) and (2.18):

$$h = \begin{bmatrix} \mathfrak{R} \\ \mathfrak{D} \end{bmatrix} = \begin{bmatrix} \|\vec{BT}\| + \|\vec{TI}\| - \|\vec{BI}\| \\ f_{\mathrm{I}} \cdot \left( \frac{\vec{v}}{c} \cdot \left( \frac{\vec{TB}}{\|\vec{TB}\|} - \frac{\vec{IT}}{\|\vec{IT}\|} \right) \right) \end{bmatrix}. \tag{4.39}$$

## 4.4.2. Tracking Algorithm

In this section, the individual steps of the tracking algorithm already mentioned above are described in more detail.

### Target estimation and Clustering

The tracking algorithm is executed when new measured values are available. The measurements always consist of sets of range, Doppler and angle information. Furthermore, the positions of the illuminators and that of the observer are assumed to be known. As elaborated in Section 4.3.3, these information are not sufficient to calculate the searched position and velocity of the target, but are sufficient for a 2-dimensional estimate.
This estimation of the target's position can be performed similar to the PA method described in Section 4.3.3. Instead of starting from Equation (4.34), the procedure is started with

$$\vec{T} = \begin{bmatrix} B_{\mathrm{x}} - R\cos\beta\sin\varphi - R\sin\beta\cos\theta\cos\varphi + A\frac{B_{\mathrm{x}}}{\|\vec{B}\|} \\ B_{\mathrm{y}} + R\cos\beta\cos\varphi - R\sin\beta\cos\theta\sin\varphi + A\frac{B_{\mathrm{y}}}{\|\vec{B}\|} \\ B_{\mathrm{z}} + \sin\beta\sin\theta + A\frac{B_{\mathrm{z}}}{\|\vec{B}\|} \end{bmatrix}. \tag{4.40}$$

The definitions from Equation (4.33) for $\theta$ and $\varphi$ are still valid and $R$ describes the distance from the observer $\vec{B}$ to the target $\vec{T}$. For all targets an altitude of $A = 1500\,\mathrm{m}$ is assumed[25]. The angle $\beta$ describes the measured bearing towards the target. This procedure yields an estimated position for each detected target.

Based on these estimates, clusters of measurements from different illuminators that potentially belong to the same target are sought next. For this purpose, the distances between the targets are compared with each other. If the distance is below a threshold, the two targets are associated with a cluster. If one of the two targets already belongs to a cluster, the second target is assigned to this cluster, as well.
After all clusters are found, the position and velocity is calculated for each one that has reached a minimum size of three measurements. First, the position of the target is determined with Equation (2.17). This is solved with the help of the GNI procedure explained in Section 4.3.3. With the position of the target now known, the Equation (2.18) can then be solved using the same procedure.
At the end of this step, for each cluster that contained more than three measurements, a starting point is obtained that could be used as a basis for a new track. In addition, after this step, estimates of the position are available for all recorded range-Doppler measurements. However, before new tracks are created, the system checks whether the measurement

[25] For the procedure described previously, a height of 0 m was assumed. The initial equation thus becomes longer by one term each, but the solving method does not change.

data can be associated with existing tracks. This means that the predicted positions of the existing tracks have to be calculated before new tracks are initialized.

### Track Prediction

After this preprocessing of the measurements, the tracks' UKFs are processed. First the new states and covariance matrices of all existing tracks are predicted using Equation (4.37). The prediction proceeds according to the Equations (2.69) - (2.71), which are described in Chapter 2.9.2.
From the estimated position and velocity of the target at the last state, the new position of the target is extrapolated. The velocity is assumed to be constant[26]. In addition, the covariance matrix of the state is updated and the process noise is applied.

### Track Update

The crucial step in tracking is updating each track with new measurements. The update itself is performed according to the Equations (2.72) - (2.79) described in Section 2.9.2. Before that, however, the measured values must be assigned to the tracks.
For this purpose, the measurement data is processed individually for each illuminator. The first step is to pre-select the measurement data that potentially belongs to the track. The values of range, bistatic velocity[27], bearing and the estimated position are compared with the predicted values of the track. If one of these values is outside a predefined threshold[28], the measurement is not associated with the track. This pre-filtering ensures that a track is not updated with implausible measurements.
Ideally, after this pre-filtering, only one measurement per illuminator remains that can be assigned to the trace. If several measurements remain, the update is calculated for all measurements individually. The results are compared with each other and the measurement that led to the smallest change of the predicted state is selected for the update. With this robust procedure, the influence of possible false assignments is minimized, while the possibility of correcting the track with correct measurements is preserved.

Before the update is applied, a "sanity check" is applied. For this purpose, it is checked whether the correction to be performed can be executed with a predefined maximum acceleration[29] $a$. If either the new position or the new velocity vector cannot be reached with the defined acceleration limit, the update is rejected. This procedure is applied sequentially for all illuminators and tracks.

### Track initialization

For the initialization of new tracks, the positions of potential new tracks obtained from the clustering of the position estimation are now compared with those of the updated existing

[26] This is due to the relatively simple model that is used here - more complex models that for example also represent the acceleration of the targets would also be possible.

[27] The bistatic velocity $v_\mathrm{b}$ is used here, since it can be defined independent from the frequency of the illuminator $f_\mathrm{I}$. Therefore, the same value can be used for all illuminators. It is given by $v_\mathrm{b} = \frac{f_\mathrm{I}+\mathfrak{D}}{f_\mathrm{I}-1} c$

[28] The following designations are defined for the threshold values: Threshold Range $c_\mathrm{R}$, threshold bistatic velocity $c_\mathrm{V}$, threshold bearing $c_\mathrm{A}$ and threshold position $c_\mathrm{D}$.

[29] This acceleration must of course be adapted to the expected targets. For a passenger aircraft or other civil aircraft a maximum acceleration in the range of $a = 10\,\mathrm{m/s^2}$ can be assumed.

tracks. If the tracks' distance is below a threshold $D_{\mathrm{Lock}}$, the track is discarded. Otherwise, a new track is initialized.
For this purpose, a new UKF is created. To initialize the track, first the position and the velocity vector are entered into the state of the UKF. In addition, the covariance matrix has to be initialized. For the implementation of tracking shown here, it has proven to be sufficient to fill this matrix with statically predefined values.
However, in further development, it would be conceivable to calculate this dynamically from the SNR of the measurements, according to Equation (4.14).

#### Track deletion

The last step in the tracking algorithm is the deletion of old tracks, i.e., tracks that have not been updated for some time and which states have therefore been available only as a prediction for a long time. The decision to delete a track is handled comparatively simply via a number of time steps $n_{\mathrm{t}}$.
If a track has not been updated with measured values within the last $n_{\mathrm{t}}$ time steps, it will be deleted. This procedure can be used to delete tracks that move out of the measuring range of the system or tracks that are created by chance ("ghost targets").

## 4.5. Performance Model

In order to be able to better assess the results of the measurements shown in the next chapter, this section will establish a model for the system's performance. The intention of this modeling is to be able to make predictions about the possible coverage area for each individual transmitter as a function of the RCS of the target.
As already described in Section 2.4, the achievable SNR of the system can be described with the bistatic radar equation, defined in Equation (2.19). For this purpose, we assume a scenario in which an omnidirectional radiator is located approximately 26 km southeast of the receiver. The transmitter broadcasts an FM radio signal with an effective bandwidth of $B_{\mathrm{eff}} = 50\,\mathrm{kHz}$ at a center frequency of 104.6 MHz, it achieves an Effective Radiated Power (ERP) of 87 kW. For the transmit antenna, a dipole characteristic is assumed, which radiation characteristic is described with Equation (3.1); it only depends on the elevation angle $\vartheta$. This results in an equivalent isotropic radiated power (EIRP) $P_{\mathrm{I,EIRP}}(\vartheta)$ that depends on the elevation angle and can be described by

$$P_{\mathrm{I,EIRP}}(\vartheta) = 1.64 \cdot P_{\mathrm{I,ERP}} \cdot \frac{\cos^2(\frac{\pi}{2}\cos(\vartheta))}{\sin^2(\vartheta)}. \tag{4.41}$$

The target is assumed to be an object with an RCS of $\sigma = 1000\,\mathrm{m}^2$, which is defined as the average bistatic RCS of a target of interest[30].
For the measurement system, the gains for the VHF array as described in Section 3.2 are assumed. Furthermore, it is assumed that the beamforming towards the target signal is panned only in the azimuth plane and is always optimally aligned. Thus, the gain of the receiving antenna can also be modeled as only dependent on the elevation angle $\vartheta$. These approximations agree with the data processing actually performed.

[30] In [90], the bistatic RCS of an Airbus A321 is measured at 117.3 MHz. It ranges from 18 dBm$^2$ to 40 dBm$^2$ with an average of 30 dBm$^2$. Similar results are shown based on a scaled model of an A380 in [91]

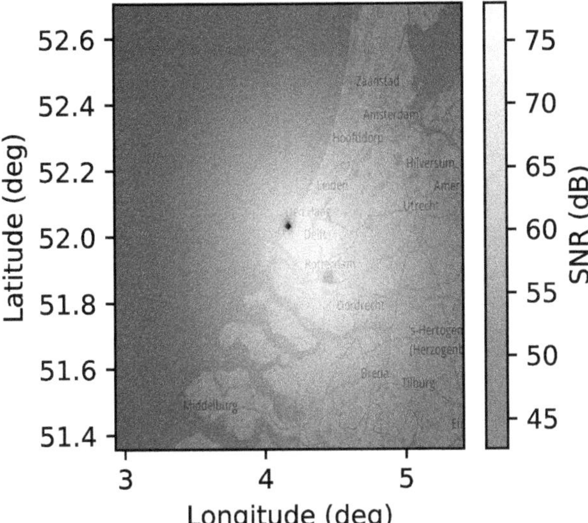


Figure 4.17.: Distribution of SNR for a target with an RCS of $1000\,\mathrm{m}^2$ at an altitude of 6000 m considering antenna characteristics. The position of the receiver is indicated by a black dot in the center of the figure. The transmitter is marked by a green square.
Map Data ©OpenStreetMap contributors [92].

With these considerations, the SNR at the receiver can be estimated using the radar equation. In addition, the process gain, which according to Equation (4.13) is composed of the bandwidth and the integration time, is taken into account.

Thus, the ratio of a target signal with respect to thermal noise and the system's noise figure can be estimated as follows:

$$\mathrm{SNR} = \frac{P_{\mathrm{I,EIRP}}(\vartheta)}{4\pi R_T^2} \cdot \sigma \cdot \frac{1}{4\pi R_R^2} \cdot \frac{G_{\mathrm{R}}(\vartheta)\lambda^2}{4\pi} \cdot \frac{1}{k_B T_0 B F} \cdot G_{\mathrm{process}}. \tag{4.42}$$

Figure 4.17 shows the result of this function for a simulated target at an altitude of 6000 m as a heatmap. The section shown covers an area[31] of about 150 km x 170 km. In comparison with the plot shown in Figure 2.4 on page 11, two differences are noticeable. First, there is a strong offset due to the process gain; second, the antenna characteristics for the high-flying target manifests as a dip in SNR around the transmitter and receiver position.
It is also noticeable that the SNR does not fall below a value of 40 dB in the entire region. Thus, if the thermal noise, or the portions of the system noise described by the noise figure $F$ are the only negative influences on the system performance, the described radar system should work very well in the entire considered area. This estimate can therefore

[31] This example is chosen deliberately and also corresponds to the practical example shown in the next chapter.

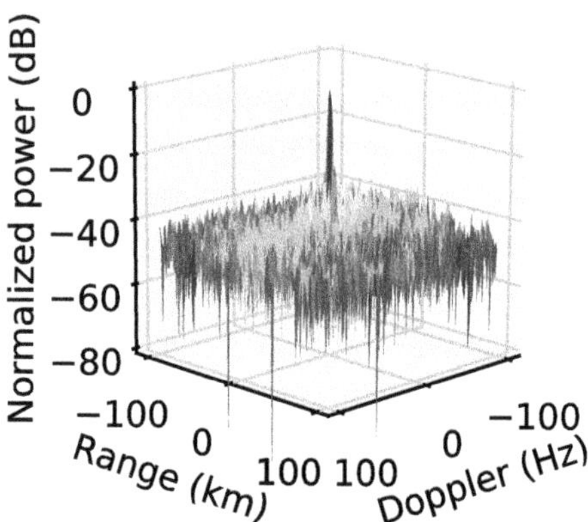


Figure 4.18.: Exemplary representation of an ambiguity function for an FM broadcasting station.

be considered a best case scenario and represents the absolute upper limit of the system's performance. In reality, however, there are some factors that have a negative impact on the performance of the system.

To understand these limitations, first the ambiguity function from Equation (2.29) is considered. In Figure 4.18 a plot of this function is shown. In this case the plot is generated from a single measurement of an FM broadcast signal from the measurement campaign described in the next chapter.

The main peak occurs for a shift of $m = 0$ range bins and $k = 0$ Doppler bins. In addition, a noise floor approximately 45 dB below the peak, fills the complete range-Doppler plane. From Section 4.2.6 it is known that the calculation of the range-Doppler matrices is done with the cross-ambiguity function in Equation (4.8). It represents a cross-correlation with a surveillance signal and is therefore closely related to the ambiguity function, which is the autocorrelation function. For signal components of the LOS illuminator signal, which are included as spurious components in the surveillance signal, a similar behavior is thus to be expected.

In other words, for signal components of the direct LOS illuminator signal, which are still present in the surveillance signal, an additional interference floor, approximately 45 dB below the "DC-peak"[32] is introduced to the system. Depending on the magnitude of this signal component, targets could be masked by this interference. To understand whether this effect is important, the magnitude of this signal component is modeled below.

First, the powers of the target signal and LOS signal at the receiver location are calculated. For now only the EIRP and the free space attenuation will be considered for the LOS signal

[32]This peak is the result of Equation (4.11) for a signal shifted neither in range nor in Doppler.

from the illuminator. This results in

$$\hat{P}_{\mathrm{I}} = P_{\mathrm{I,EIRP}}(\vartheta)\left(\frac{\lambda}{4\pi R_{\mathrm{IB}}}\right)^2 \tag{4.43}$$

with $R_{IB} = \|\vec{BI}\|$, which is the distance between the illuminator and the receiver. For the signal reflected at the target, in addition to the propagation between transmitter and target, the RCS and the distance from the target to the receive antenna are relevant as well. This results in

$$\hat{P}_{\mathrm{T}} = \frac{P_{\mathrm{I,EIRP}}(\vartheta)}{4\pi R_{\mathrm{T}}^2} \cdot \sigma \cdot \frac{1}{4\pi R_{\mathrm{B}}^2} \cdot \frac{\lambda^2}{4\pi}. \tag{4.44}$$

With these equations, an example consideration of the relationship between the target signal and the illuminator signal at the receiver can be made. To do this, the illuminator, target, and receiver are assumed to be in the same horizontal plane and the target and receiver are assumed to be at the same distance from the illuminator. It is also assumed that the target is at a distance of 10 km from the receiver. Under these assumptions, the results are $\hat{P}_{\mathrm{I}} \approx 11\mathrm{e}{-3}\,\mathrm{W} \hat{\approx} -20\,\mathrm{dBm}$ and $\hat{P}_{\mathrm{T}} \approx 8.75\mathrm{e}{-9}\,\mathrm{W} \hat{\approx} -80\,\mathrm{dBm}$. This example shows that due to the double distance dependence of the target signal, it will almost always appear to be orders of magnitude weaker than the direct illuminator signal. The interference from the LOS illuminator signal described by its ambiguity function would mask all target signals at this ratio of powers.
In passive radar systems, there are therefore several measures to minimize the contribution of the direct illuminator signal to the surveillance signal. The improvements to the surveillance signal by these measures must be considered for the model. The relevant steps and their effects are shown schematically in Figure 4.19 and are described in the following paragraphs.

In this system, the first step in separating the reference channel from the surveillance channel is the use of digital beamforming. For the purpose of modeling it is assumed that the reference beam is calculated in the azimuth plane directly towards the illuminator. The power of the sampled signals therefore results from the signal power and the respective gains from the beamformer[33].

This means that in the surveillance signal the power of the target is received with maximum gain, while the power of the transmitter is received only with a side lobe. Assuming that the beam pattern is symmetrical, the same applies vice versa for the reference signal. For the model, antenna losses are neglected, and $G(\theta, \varphi) = D(\theta, \varphi)$ applies. The different gains can be calculated with Equation (2.42). Based on these various gains and the powers of the signals at the receiver position already calculated above, the power of each signal component after the beamforming can be set up as follows[34]

[33]The influence of additional amplifiers or losses on cables, etc. are represented by the noise figure and since they may be assumed to be identical for all paths, they can be neglected in the modeling.

[34]The following consideration assumes a single measurement for which the azimuth ($\vartheta$) and elevation ($\varphi$) angles to the target and illuminator, respectively, are constant. The symmetrical structure of the array used here means that the main and side levels are the same in each case, but this does not necessarily apply to more complex arrays.

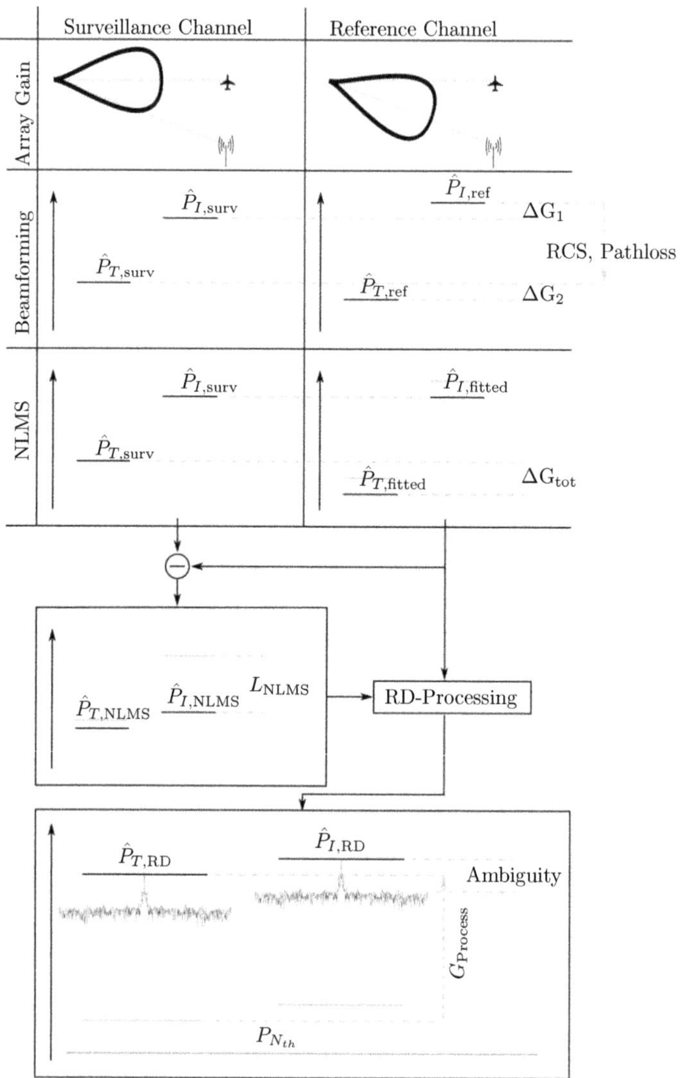


Figure 4.19.: Schematic overview of the performance model.

$$\begin{aligned}\hat{P}_{\mathrm{T,surv}} &= G_{\mathrm{surv,main}}\hat{P}_{\mathrm{T}}\\ \hat{P}_{\mathrm{I,surv}} &= G_{\mathrm{surv,side}}\hat{P}_{\mathrm{I}}\\ \hat{P}_{\mathrm{I,ref}} &= G_{\mathrm{ref,main}}\hat{P}_{\mathrm{I}}\\ \hat{P}_{\mathrm{T,ref}} &= G_{\mathrm{ref,side}}\hat{P}_{\mathrm{T}}\end{aligned} \tag{4.45}$$

After beamforming, the power of the surveillance signal is

$$\mathrm{Surv} = \hat{P}_{\mathrm{T,surv}} + \hat{P}_{\mathrm{I,surv}} \tag{4.46}$$

and that of the reference signal is

$$\mathrm{Ref} = \hat{P}_{\mathrm{I,ref}} + \hat{P}_{\mathrm{T,ref}}. \tag{4.47}$$

Next, the behavior of the NLMS algorithm described in Section 4.2.5 is considered. This algorithm aims to suppress the illumination signal contained in the surveillance channel. For this purpose, the share of the illumination signal contained in the surveillance channel is modeled from the reference signal. The modeled signal is then subtracted from the surveillance channel. For the modeling of the process, this procedure is represented by the scaling of the reference signal by a factor $a$. Moreover, it will be taken into account that the real suppression is not ideal, therefore the factor $L_{\mathrm{NLMS}}$ is additionally introduced, which is used to describe the achieved suppression of the algorithm. It results in

$$\mathrm{NLMS} = \mathrm{Surv} - a \cdot \mathrm{Ref} \cdot \left(1 - \frac{1}{L_{\mathrm{NLMS}}}\right). \tag{4.48}$$

From the consideration that the NLMS aims to emulate the illuminator signal, the following condition for $a$ can be defined:

$$\begin{aligned}\hat{P}_{\mathrm{I,surv}} &\overset{!}{=} a \cdot \hat{P}_{\mathrm{I,ref}}\\ G_{\mathrm{surv,side}}\hat{P}_{\mathrm{I}} &= a \cdot G_{\mathrm{ref,main}}\hat{P}_{\mathrm{I}}\\ a &= \frac{G_{\mathrm{surv,side}}}{G_{\mathrm{ref,main}}}\end{aligned} \tag{4.49}$$

This condition for $a$ inserted into Equation (4.48) yields:

$$\mathrm{NLMS} = \underbrace{\mathrm{Surv}}_{\mathrm{A}} \underbrace{- \frac{G_{\mathrm{surv,side}}}{G_{\mathrm{ref,main}}} \cdot \mathrm{Ref}}_{\mathrm{B}} + \underbrace{\frac{G_{\mathrm{surv,side}}}{G_{\mathrm{ref,main}}} \frac{\mathrm{Ref}}{L_{\mathrm{NLMS}}}}_{\mathrm{C}} \tag{4.50}$$

The three summands are considered individually and the signals defined in Equations (4.45) - (4.47) are substituted:

$$
\begin{aligned}
\mathrm{A} &= G_{\text{surv,main}} \hat{P}_{\mathrm{T}} + G_{\text{surv,side}} \hat{P}_{\mathrm{I}} \\
\mathrm{B} &= \frac{G_{\text{surv,side}} G_{\text{ref,side}}}{G_{\text{ref,main}}} \hat{P}_{\mathrm{T}} + G_{\text{surv,side}} \hat{P}_{\mathrm{I}} \\
\mathrm{C} &= \frac{G_{\text{surv,side}} G_{\text{ref,side}}}{G_{\text{ref,main}} L_{\text{NLMS}}} \hat{P}_{\mathrm{T}} + \frac{G_{\text{surv,side}}}{L_{\text{NLMS}}} \hat{P}_{\mathrm{I}}
\end{aligned}
$$

Then the signal is partitioned into the components of the target signal and the illuminator signal:

$$
\begin{aligned}
\mathrm{NLMS} = G_{\text{surv,main}} \left(1 - \frac{G_{\text{surv,side}}}{G_{\text{ref,main}}} \frac{G_{\text{ref,side}}}{G_{\text{surv,main}}} \left(1 - \frac{1}{L_{\text{NLMS}}}\right)\right) \hat{P}_{\mathrm{T}} \\
+ \left(\frac{G_{\text{surv,side}}}{L_{\text{NLMS}}}\right) \hat{P}_{\mathrm{I}}
\end{aligned}
\tag{4.51}
$$

From this equation it follows that the ratio of the illuminator signal in the surveillance channel becomes smaller by the factor $L_{\text{NLMS}}$:

$$
\hat{P}_{I,\text{NLMS}} = \frac{\hat{P}_{I,\text{surv}}}{L_{\text{NLMS}}}. \tag{4.52}
$$

As an undesirable side effect, however, the power of the target signal is also reduced. The remaining target signal after the NLMS algorithm can be described with

$$
\hat{P}_{T,\text{NLMS}} = \hat{P}_{T,\text{surv}} \cdot \left(1 - \frac{G_{\text{surv,side}}}{G_{\text{ref,main}}} \frac{G_{\text{ref,side}}}{G_{\text{surv,main}}} \left(1 - \frac{1}{L_{\text{NLMS}}}\right)\right). \tag{4.53}
$$

Assuming that the NLMS achieves good suppression ($L_{\text{NLMS}} >> 1$), this term can be simplified to:

$$
\hat{P}_{T,\text{NLMS}} \approx \hat{P}_{T,\text{surv}} \cdot \left(1 - \frac{G_{\text{surv,side}}}{G_{\text{ref,main}}} \frac{G_{\text{ref,side}}}{G_{\text{surv,main}}}\right) \tag{4.54}
$$

The two fractions that remain describe the relationships between the illuminator and target signal components between the reference and surveillance channels (see Figure 4.19):

$$
\Delta G_1 = \frac{\hat{P}_{\mathrm{I,ref}}}{\hat{P}_{\mathrm{I,surv}}} = \frac{G_{\text{surv,side}}}{G_{\text{ref,main}}}, \tag{4.55}
$$

$$
\Delta G_2 = \frac{\hat{P}_{\mathrm{T,ref}}}{\hat{P}_{\mathrm{T,surv}}} = \frac{G_{\text{ref,side}}}{G_{\text{surv,main}}}, \tag{4.56}
$$

$$
\Delta G_{\text{tot}} = \Delta G_1 \cdot \Delta G_2. \tag{4.57}
$$

Consequently, this results in

$$
\hat{P}_{T,\text{NLMS}} \approx \hat{P}_{T,\text{surv}} \cdot (1 - \Delta G_{\text{tot}}). \tag{4.58}
$$

To model this, it is divided into two steps. In step one the illuminator power of the reference channel is aligned with the surveillance channel. As can be seen in Figure 4.19.

By definition, the antenna gains are greater than zero and the main components are larger or equal to those of the side lobes. Therefore, it follows that $0 < \Delta G_1 \leq 1$ and $0 < \Delta G_2 \leq 1$. Thus $0 < \Delta G_{\text{tot}} \leq 1$ is also valid.
From this consideration it follows that in the best case the power of the target signal is not affected by the NLMS. In the worst case, i.e. when the four gains described in Equation (4.45) are the same, the target signal is completely suppressed.

To complete the model of the system, the last step of the data processing is the creation of the range-Doppler matrices by cross-correlation. This process can be modeled by applying the process gain described by Equation (4.13). The final power of the target signal can be modeled as

$$\hat{P}_{T,\text{RD}} = G_{\text{process}} \hat{P}_{T,\text{NLMS}}. \tag{4.59}$$

Now that the signal power has been modeled, the next step is to determine the noise as well as the interference components of the signal.
The noise floor is composed of the thermal noise, which can be described with

$$N_{\text{thermal}} = k_{\text{B}} T_0 B F. \tag{4.60}$$

Where $k_{\text{B}}$ is the Boltzmann constant, $T_0$ is the assumed noise temperature, $B$ is the system bandwidth, and $F$ is the system noise figure. As described above, the side lobes of the ambiguity function are now added as an additional interference. The power of this component is defined at a fixed offset $\Delta$ambiguity from the illuminator signal remaining in the NLMS signal:

$$N_{\text{ambiguity}} = \frac{G_{\text{process}} \hat{P}_{I,\text{NLMS}}}{\Delta \text{ambiguity}}. \tag{4.61}$$

With these expressions the signal-to-noise and interference ratio (SNIR) can be described with

$$\text{SNIR} = \frac{\hat{P}_{T,\text{RD}}}{N_{\text{thermal}} + N_{\text{ambiguity}}}. \tag{4.62}$$

With a few assumptions for the system properties, this function can now be plotted: The system operates with an integration time of $t_{\text{i}} = 670\,\text{ms}$ and an effective bandwidth of $B_{\text{eff}} = 50\,\text{kHz}$ can be assumed. The resulting process gain is therefore, $G_{\text{process}} = 33500 \approx 45\,\text{dB}$. An equivalent noise temperature of $T_0 = 290\,\text{K}$ and a system bandwidth of $B = 200\,\text{kHz}$ is assumed. The system noise figure is assumed to be $F = 16\,\text{dB}$[35].

In Figure 4.20 the results are plotted for the illuminator described at the beginning of this section. For the calculation of the gains due to beamforming, the positions of the antennas are offset with a normally distributed random number ($\sigma = 0.03\,\text{m}^2$). This is done to relax the zeros in the antenna pattern that are still prominently featured in the plot.
It can be clearly seen that the range of the system with this model is determined by the interference and not by the thermal noise, the maximum SNIR achieved is about 45 dB, which is well below the minimum SNR due to thermal noise, that is shown in Figure 4.17. The plot also clearly shows that in direct bearing towards the illuminator the SNIR drops to zero. From the consideration that the SNIR in this scenario depends significantly on the $\hat{P}_T/\hat{P}_I$ ratio, it can also be concluded that the system performance depends more on the

[35] In Chapter 3.3 a significantly better noise figure was determined, but this was measured under laboratory conditions, at the maximum pre-amplification of the signal in the SDR receiver. Under real conditions, however, this cannot be used because the direct LOS signals of the transmitters are received with too much power, so the preamplification was reduced and, thus, the noise figure was larger.

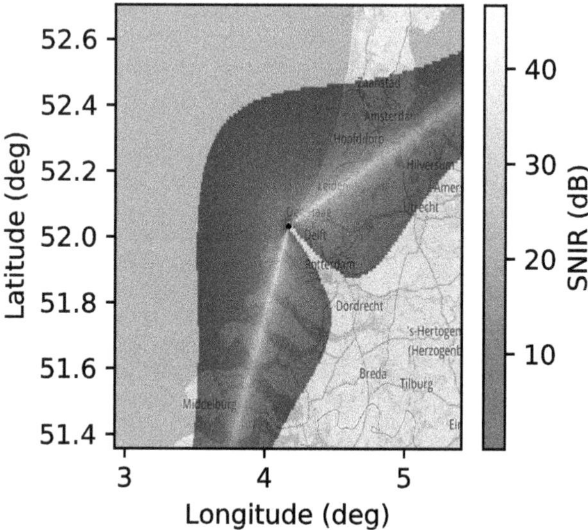


Figure 4.20.: Distribution of SNIR for a target with an RCS of $1000\,\mathrm{m}^2$ at an altitude of 6000 m considering antenna characteristics. In the areas without overlaid SNIR plot, the signal power is lower than that of the noise/interference. Map Data ©OpenStreetMap contributors [92].

transmitter's location and less on the transmitting power[36]. These considerations apply to a single target. If there is more than one target involved, the ambiguity functions of all targets will contribute to the interference. These individual contributions would then have to be weighted by the antenna characteristics as well. Thus, even if an ideal function of the NLMS algorithm is assumed, the system performance would not depend on thermal noise alone. With these results a model for the system performance is obtained, which can now be compared with actual measured values in the next chapter.

[36]As long as the received powers including the process gain are clearly above the noise floor.

# 5. Performance in a Real-World Scenario

In this chapter, the performance of the system is demonstrated using real measured data. As already described in Section 4.3.5, measurements were carried out in the Netherlands, near The Hague. To be able to verify the results, ADS-B data were also recorded as ground truth. The recorded ADS-B was later supplemented with ADS-B recordings from the OpenSky database [93]. In the following sections a dataset of about 6 minutes in length recorded on August 21, 2019 is discussed. For the purpose of system development, the non-interpolated range Doppler matrices were recorded and used as a basis for further data processing. The signal processing described in Section 4.2 was performed offline for the results shown here, starting with the CFAR algorithm; however, online processing of such data is of course possible in principle.

## 5.1. Scenario Description

A total of thirteen transmitters at five different locations are used for the measurements. The characteristics of the transmitters shown here are taken from the open database "fm-scan.org" [53]. The key data of all used transmitting stations is summarized in Table 5.1. The distance $d$ to the receiver is provided and as a rough measure of the illuminator's performance, the sum of detections during the measurement period is also listed here.
Figures 5.2 and 5.3 show the measurements for each transmitter collected during the campaign. In the illustrations each transmitter position is marked with a colored square. The positions corresponding to a detected target are displayed in the same color as the associated illuminator with star markers. They are approximated with the PA method, assuming a target altitude of 6000 m. The position of the receiving system is marked with a black dot in the center of the plots.
The targets are "targets of opportunity" and consist mostly of commercial flights. Most of the targets in the measurement period sent ADS-B information, which can be used as ground truth information. This information is recorded and interpolated as described in Section 4.3.2. The ground truth obtained in this way is marked with white crosses in the figures.
Since it turned out that the range of the system exceeded the range of the recorded ADS-B reference, further reference data from the open database "OpenSky Network" is also included [93]. These reference tracks are drawn as white lines and the respective starting point is marked with a white diamond.[1]
Also plotted with yellow markers in the northern part of the figure are the positions of three offshore wind farms [94], which have been identified as possible sources of Doppler-affected clutter. As a reference for the expected system range, the plots are complemented with the model described in Section 4.5. The data from the model is presented as an overlaid heatmap of the SNIR and is based on a target with an RCS of $1000\,\mathrm{m}^2$ at an altitude of 6000 m.

[1]The start points of the tracks are related to their first sighting in the measurement period, i.e. the markers do not necessarily represent the same point in time.

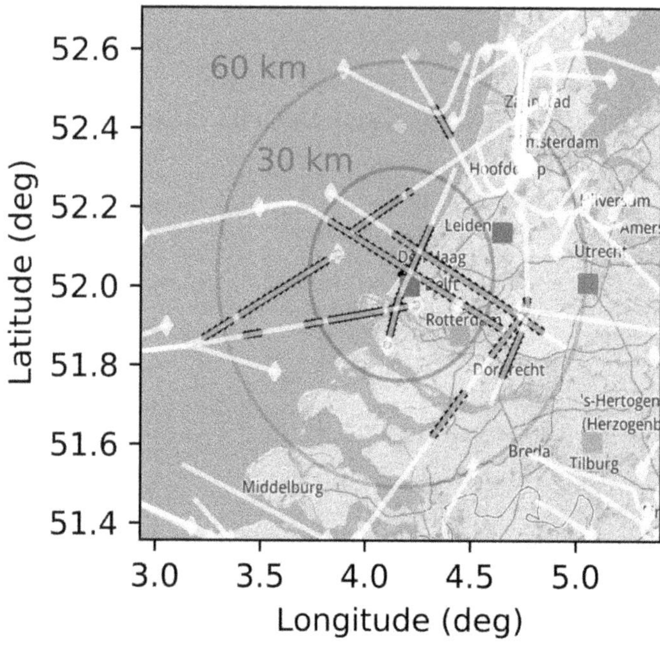


Figure 5.1.: Overview of the measurement scenario. The black crosses mark ADS-B measurements recorded at the radar's position. The white lines mark the tracks from the OpenSky data set, with the white diamond marking the starting point. The colored squares mark the illuminator locations.
Map Data ©OpenStreetMap contributors [92].

The visualization of the SNIR starts only from a lower limit of 7.5 dB, which corresponds to the applied lower threshold of the CFAR algorithm. For the plot, the maximum values are also clipped at 30 dB.

## 5.2. Illuminator Performance

This section describes and compares the results for each illuminator. First, some common features of the modeled SNIR can be found in the Figures 5.2 and 5.3. Initially, it is noticeable that in the areas where the illuminator and target are within the main lobe of the antenna, the SNIR does not allow any detections. The system is therefore blind in the direction of the illuminator. The best performance is found in two zones almost orthogonal to the baseline between illuminator and receiver. These occur when the illuminator is in a null of the antenna pattern while the target is in the main lobe. Since the position of this null is frequency dependent, the angle at which these zones lie with respect to the baseline is also variable. This behavior is particularly evident in Figure 5.3 where the results of

Table 5.1.: Overview of the illuminators used. The color codes behind the illuminator numbers are assigned to the illuminator positions. They are used in the following figures to display the measurement data. The power $P$ is ERP.

| Illuminator | $f$ / MHz | Lat / deg | Lon / deg | $P$ / kW | $d$ / km | Detections |
|---|---|---|---|---|---|---|
| 00 | 87.6 | 52.00147 | 4.21059 | 0.79 | 4.6 | 17 |
| 01 | 95.2 | 52.13711 | 4.64647 | 25 | 35.0 | 322 |
| 02 | 96.8 | 52.01005 | 5.05359 | 72 | 61.0 | 157 |
| 03 | 102.7 | 51.87569 | 4.44848 | 100 | 26.0 | 90 |
| 04 | 98.9 | 52.01005 | 5.05359 | 72 | 61.0 | 46 |
| 05 | 92.6 | 52.01005 | 5.05359 | 72 | 61.0 | 128 |
| 06 | 98.2 | 51.60825 | 5.07698 | 55 | 78.5 | 223 |
| 07 | 103.8 | 51.87569 | 4.44848 | 20 | 26.0 | 115 |
| 08 | 104.6 | 51.87569 | 4.44848 | 87 | 26.0 | 90 |
| 09 | 101.5 | 51.87569 | 4.44848 | 8.3 | 26.0 | 63 |
| 10 | 100.4 | 51.87569 | 4.44848 | 10 | 26.0 | 15 |
| 11 | 97.6 | 51.87569 | 4.44848 | 18 | 26.0 | 121 |
| 12 | 94.7 | 51.87569 | 4.44848 | 13 | 26.0 | 84 |

six transmitters at the same location are shown. In all plots there is also an attenuation of the SNIR in the area of the receiver. This is a result of the chosen cut at an altitude of 6000 m and again is a result of the antenna pattern of the array, which is composed of dipole antennas.

Figures 5.2(a) - 5.2(c) show the results for the three illuminator positions for which only one transmitted frequency is used. Figure 5.2(a) shows the results for illuminator "00". With an ERP of $P_{\mathrm{I}} = 790\,\mathrm{W}$, it is the weakest transmitter that is used. It is located at a distance of approximately 4.6 km, and is thereby closest to the receiver location.
Due to the closeness to the receiver position, the model shows a comparatively small area with usable SNIR. Clearly visible are the two elongated areas that run almost orthogonal to the baseline between the transmitter and receiver and are created by the minima in the antenna pattern. Most of the targets shown correspond very well with the reference tracks and are mostly in the range of the highest predicted SNIR.
During the measurement period, a total of 17 potential targets could be detected with this illuminator.

Figure 5.2(b) shows the results for the northernmost illuminator "01". It operates with a ERP transmit power of $P_{\mathrm{I}} = 25\,\mathrm{kW}$ and is located at a distance of approximately 35 km. The modeled SNIR shows a rather large area with a suitable SNIR.
With a total of 322 detections, this illuminator collected the most data within the measurement period. The first thing to notice is that also with this illuminator, most of the detections seem to match well with the reference tracks and the model.
Two things stand out in the distribution of the approximated target positions. First, there is a noticeable clustering of detections in the area of the wind farms. Second, it is noticeable that there are significantly fewer detections in the direction towards the transmitter. This is an indication that the described model for the SNIR accurately represents the system properties.

Figure 5.2(c) shows the results of the southeastern illuminator 06, which is the most remote illuminator in the system at a distance of approximately 78.5 km. The illuminator

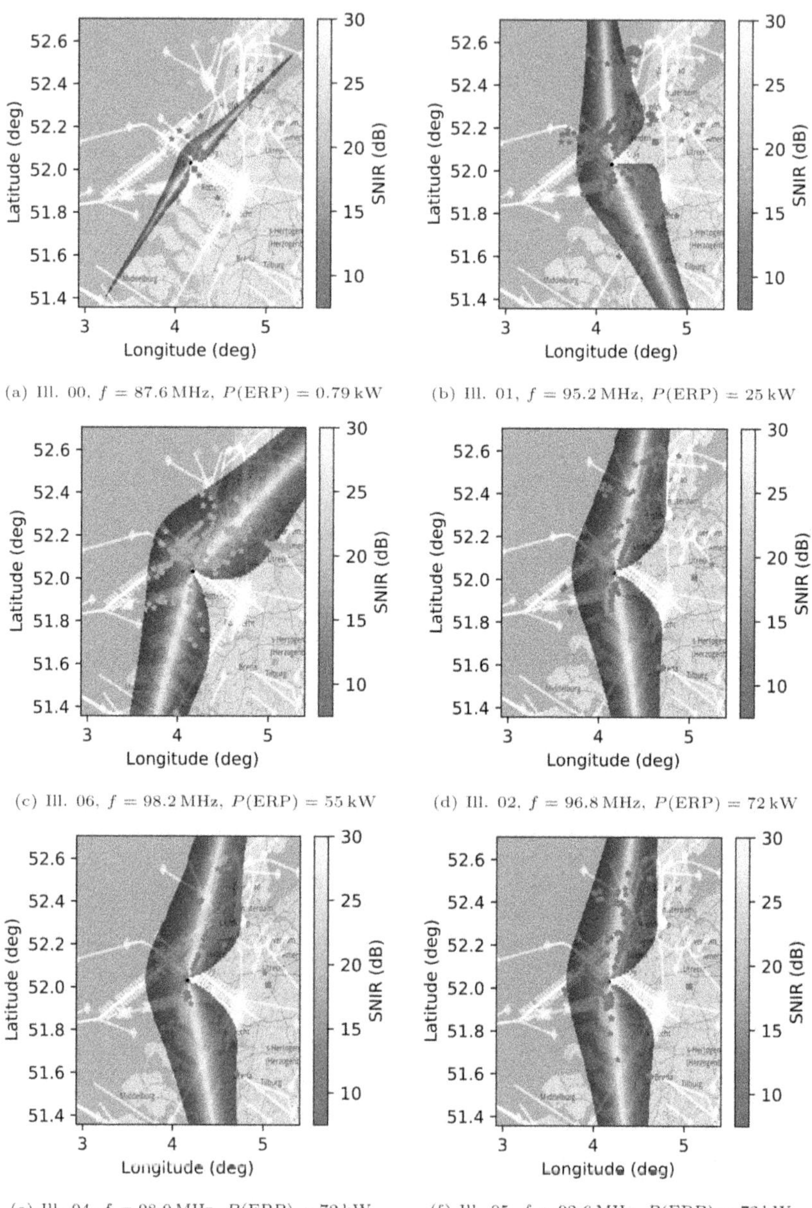


(a) Ill. 00, $f = 87.6\,\text{MHz}$, $P(\text{ERP}) = 0.79\,\text{kW}$

(b) Ill. 01, $f = 95.2\,\text{MHz}$, $P(\text{ERP}) = 25\,\text{kW}$

(c) Ill. 06, $f = 98.2\,\text{MHz}$, $P(\text{ERP}) = 55\,\text{kW}$

(d) Ill. 02, $f = 96.8\,\text{MHz}$, $P(\text{ERP}) = 72\,\text{kW}$

(e) Ill. 04, $f = 98.9\,\text{MHz}$, $P(\text{ERP}) = 72\,\text{kW}$

(f) Ill. 05, $f = 92.6\,\text{MHz}$, $P(\text{ERP}) = 72\,\text{kW}$

Figure 5.2.: Estimated positions of measurements and modeled SNIR for different illuminators. Additionally, the reference tracks are shown in white.
Map Data ©OpenStreetMap contributors [92].

operates with a transmit power of $P_\mathrm{I} = 55\,\mathrm{kW}$ and 223 detections are recorded during the measurement period.
It is also clear that the detection probability seems to be significantly lower in the immediate direction of the transmitter, which again agrees well with the predictions of the model. The targets found for this illuminator also generally fit very well with the reference tracks and the model.
However, it is noticeable that for the track flying directly over the receiver, no detections are measured in an area stretching a couple of kilometers to the northwest of the receiver, although very good performance is predicted for this area. This could be an indication of much stronger selectivity in elevation than the modeling predicts, or that the bistatic RCS of the target dropped due to an unfavorable geometry of the scenario.

Figures 5.2(d) - 5.2(f) show the results for illuminators 02, 04, and 05, which are all located at the same site, about 61 km west of the receiver. The three illuminators from this location all transmit with the same transmit power of $P_\mathrm{I} = 72\,\mathrm{kW}$, therefore all have about the same distribution of the modeled SNIR[2]. According to the model, the three transmitters should provide comparable results and detect at least approximately the same targets.
However, if the numbers of targets found (157 / 46 / 128) are compared, there is a clear difference between the individual stations. Moreover, the distribution of the detected targets is approximately similar, which speaks against different radiation characteristics of the transmitters.
Illuminator 05 seems to detect the same targets as Illuminator 04 and moreover targets in larger distances. The same observation applies to illuminator 02, which seems to have the largest range. Since all measurements are taken simultaneously and with the same geometry of the scenario, this phenomenon cannot be explained by the high dynamics in the RCS of the targets either.
A possible explanation could therefore be the influence of the effective bandwidth, which has already been described extensively by Griffith [56, 57]. As a matter of fact, it could be observed that illuminator 04 broadcasted a program with very large speech content during the measurement period, while illuminators 02 and 05 broadcasted a music program.
For the three illuminators, clusters are also observed in the area of the wind turbines to the north. It is also noticeable for these illuminators that no detections are found in the area directly northwest of the receiver. Which again supports the theory of a deviation of the real antenna pattern with respect to the modeling.

Figure 5.3 shows the results for six out of the seven illuminators at the remaining fifth site. The location is approximately 26 km southeast of the receiver. The transmit powers at this site range from $P_\mathrm{I} = 8.3\,\mathrm{kW}$ to $P_\mathrm{I} = 100\,\mathrm{kW}$. The example of these transmitters shows well that the model for the SNIR here is independent of the respective transmitting power. The reason for this behavior is that the range here is limited only by the interference component of the SNIR, which of course grows proportional to the transmit power. Comparing the number of targets found, there is also no clear correlation to the transmission power.
Comparing the plots with each other, only a difference on account of the different frequencies can be seen, for which a minimally different array pattern results in each case. The detections found here again correspond well with the model and the reference tracks.
The effective bandwidth of the transmitters is probably again the decisive difference between the illuminators. As with illuminators 02, 04, 05, and 06, it is again noticeable that no detections are generated in the area a couple of kilometers northwest of the receiver. This again indicates that the modeled antenna parameters do not yet reflect the true conditions

[2] On closer inspection, the minimal differences due to the different frequencies can be seen in the modeling.

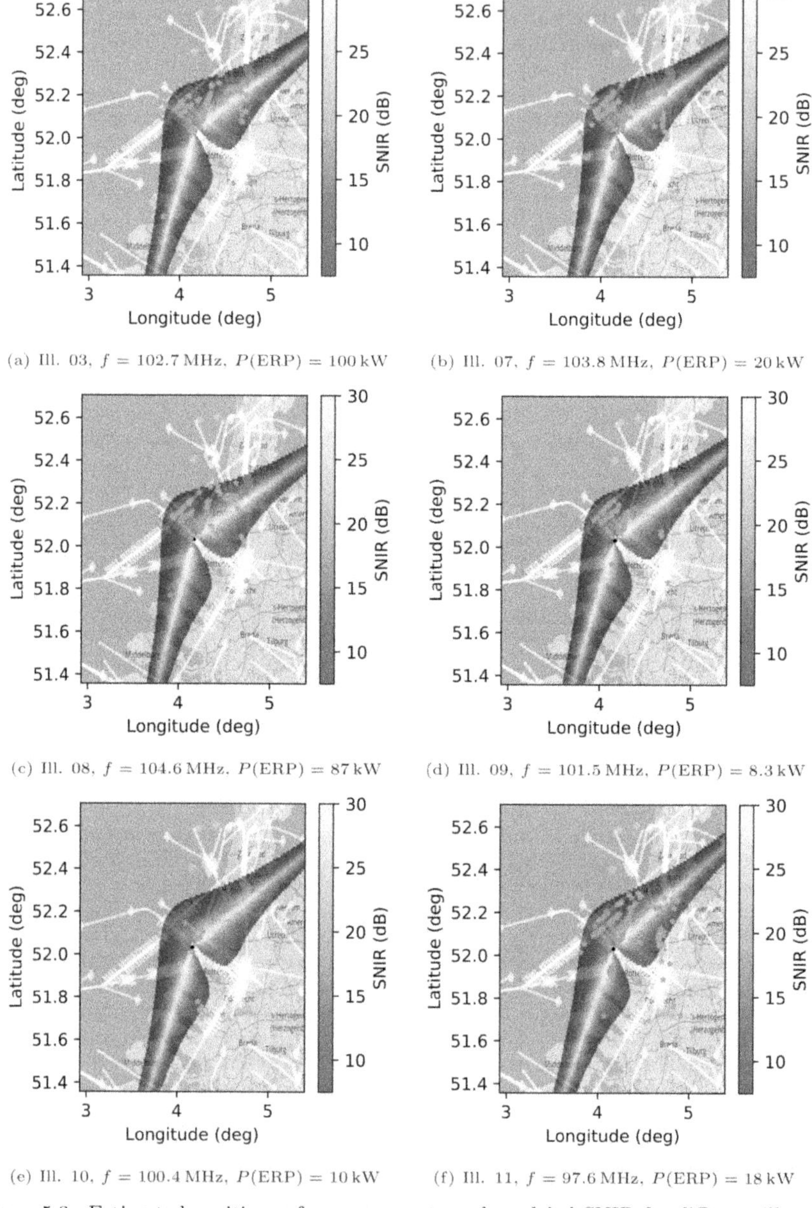


(a) Ill. 03, $f = 102.7\,\mathrm{MHz}$, $P(\mathrm{ERP}) = 100\,\mathrm{kW}$

(b) Ill. 07, $f = 103.8\,\mathrm{MHz}$, $P(\mathrm{ERP}) = 20\,\mathrm{kW}$

(c) Ill. 08, $f = 104.6\,\mathrm{MHz}$, $P(\mathrm{ERP}) = 87\,\mathrm{kW}$

(d) Ill. 09, $f = 101.5\,\mathrm{MHz}$, $P(\mathrm{ERP}) = 8.3\,\mathrm{kW}$

(e) Ill. 10, $f = 100.4\,\mathrm{MHz}$, $P(\mathrm{ERP}) = 10\,\mathrm{kW}$

(f) Ill. 11, $f = 97.6\,\mathrm{MHz}$, $P(\mathrm{ERP}) = 18\,\mathrm{kW}$

Figure 5.3.: Estimated positions of measurements and modeled SNIR for different illuminators. Additionally, the reference tracks are shown in white.
Map Data ©OpenStreetMap contributors [92].

sufficiently. By re-observing the behavior for this illuminator location, a significant decrease in bistatic RCS due to the geometry of the scenario becomes less likely.

As described in Section 4.4.2, at least three independent measurements are required to start a new track. Figure 5.4 shows the region where this condition is met for the assumed target with a heatmap of the SNIR. From all available illuminators, the third strongest SNIR is picked for each coordinate, i.e. the minimum SNIR available for track generation. For the visulization of the Heatmap a lower limit of 7.5 dB is used, which is the threshold of the used CFAR algorithm. The displayed data corresponds to an altitude cut at 6000 m. The ADS-B reference from the OpneSky database tracks are again plotted with white lines with a white diamond representing their starting position and the measured ADS-B references are marked with white crosses. The black diamonds mark all clusters consisting of at least three measurements where track generation would be possible[3]. This plot shows that the generation of tracks is possible in a kidney-shaped area around the receiver. Apart from this large, connected area, there is only one very narrow area in the southeast direction in which the conditions are fulfilled. The clusters found lie predominantly in the areas predicted by the model and also arise mostly in areas in which reference tracks are found. There is a total of three track segments that lie in the area where the SNIR is modeled as usable, for which no clusters could be formed. Two of these are in the zone directly northwest of the receiver and match the previously described absent measurements in this area, possibly indicating flaws in the model used for the antenna pattern.
The third area is observable for a track that starts south of the receiver and has a western heading. While detections are observed in this area for three different sites, it appears that the conditions for clustering are not met. This is possibly an indication for errors in the angle determination, which would lead to major deviations in the approximated position.

## 5.3. Tracking Results

To merge the individual illuminator measurements into complete tracks, the tracking algorithm described in Section 4.4 is used. The parameters used for the tracking presented in this section, are determined empirically. Estimated and constant variances describing $\boldsymbol{P}$, $\boldsymbol{R}$, and $\boldsymbol{Q}$ are used for initialization and update of the Kalman filters, as previously described in Section 4.4.2.
These might also be computed dynamically in the next iteration of the system using Equation (4.14), which would require the determination of the respective factors $k_M$.

However, it turns out that the UKFs also work sufficiently with the constant parameters and the performance of the system depends predominantly on the parameters for track management and measurement to track association. The parameters used for the UKFs are shown in Table 5.2.

The settings used for the assignment of the measurements and the general track management are summarized in Table 5.3. Figure 5.5 shows an overview of all tracks found during the measurement period. These are drawn with black lines and the starting points of the tracks are marked with a black diamond. The first five tracks are also marked with a red number; they will be discussed in further detail later on.

For orientation, two green concentric circles with a radius of 30 km respectively 60 km around the receiver location are plotted. The best system performance is achieved within

[3] Note, that only Clusters that could not be associated to an existing track started a new track.

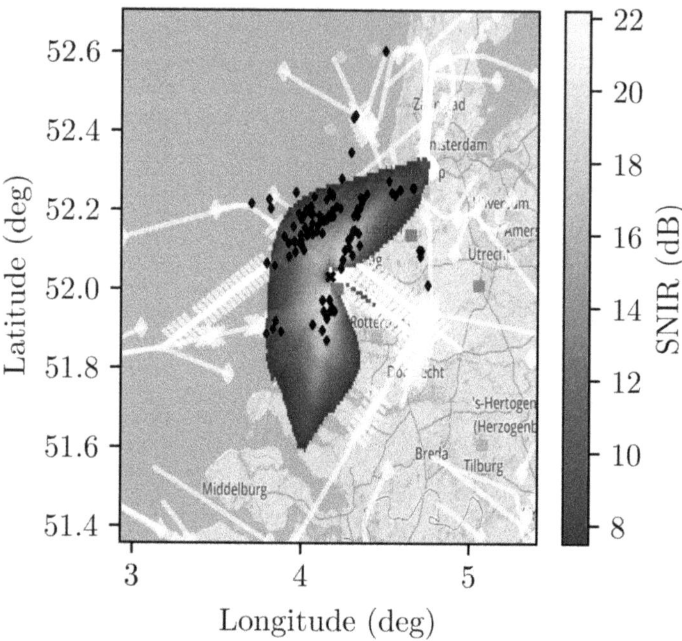


Figure 5.4.: Modeling of the area in which tracks can be started as a heatmap of the minimum available SNIR. The black diamonds are the actually identified possible starting positions for tracks found during the measurement shown.
Map Data ©OpenStreetMap contributors [92].

the inner circle. Here, for all reference targets a track could be generated and maintained over a longer period of time. In addition, even some tracks are identified that could not be matched to the available reference data.

In the Figures 5.7 - 5.11 the first five tracks are shown individually. In order to be able to assign the tracks to a reference, the reference tracks (white) are displayed in the same time window as the track under consideration (black). The starting points are again marked with diamonds. In addition, the measured values associated with the track are displayed with colored dots. The color of the dot corresponds to the location of the associated illuminator, the estimated position is again the result of the PA method.

The line from the approximated position to the track marks the position at which the track is updated with the measurement. This shows the partly large differences between the approximated and the actual positions of the targets. In the following, the results of the first five tracks are considered in more detail.

As can be seen in Figures 5.7 - 5.10, tracks 1,2,3, and 5 could be clearly assigned to reference

Table 5.2.: Parameters with which the UKFs are initialized.

| | |
|---|---|
| $\boldsymbol{P}$ | $\begin{bmatrix} 0.25\,\mathrm{km}^2 & 0 & 0 & 0 & 0 & 0 \\ 0 & 100\,\mathrm{m}^2/\mathrm{s}^2 & 0 & 0 & 0 & 0 \\ 0 & 0 & 0.25\,\mathrm{km}^2 & 0 & 0 & 0 \\ 0 & 0 & 0 & 100\,\mathrm{m}^2/\mathrm{s}^2 & 0 & 0 \\ 0 & 0 & 0 & 0 & 0.25\,\mathrm{km}^2 & 0 \\ 0 & 0 & 0 & 0 & 0 & 100\,\mathrm{m}^2/\mathrm{s}^2 \end{bmatrix}$ |
| $\boldsymbol{R}$ | $\begin{bmatrix} 0.25\,\mathrm{km}^2 & 0 \\ 0 & 4\,\mathrm{Hz}^2 \end{bmatrix}$ |
| $\boldsymbol{Q}$ | $\begin{bmatrix} 50\,\mathrm{m}^2/\mathrm{s}^2\delta t^2 & 50\,\mathrm{m}^2/\mathrm{s}^2\delta t & 0 & 0 & 0 & 0 \\ 50\,\mathrm{m}^2/\mathrm{s}^2\delta t & 50\,\mathrm{m}^2/\mathrm{s}^2 & 0 & 0 & 0 & 0 \\ 0 & 0 & 50\,\mathrm{m}^2/\mathrm{s}^2\delta t^2 & 50\,\mathrm{m}^2/\mathrm{s}^2\delta t & 0 & 0 \\ 0 & 0 & 50\,\mathrm{m}^2/\mathrm{s}^2\delta t & 50\,\mathrm{m}^2/\mathrm{s}^2 & 0 & 0 \\ 0 & 0 & 0 & 0 & 50\,\mathrm{m}^2/\mathrm{s}^2\delta t^2 & 50\,\mathrm{m}^2/\mathrm{s}^2\delta t \\ 0 & 0 & 0 & 0 & 50\,\mathrm{m}^2/\mathrm{s}^2\delta t & 50\,\mathrm{m}^2/\mathrm{s}^2 \end{bmatrix}$ |

Table 5.3.: Parameters with which the assignment of the measured values to the tracks is configured.

| Parameter | Value |
|---|---|
| $c_{\mathrm{R}}$ | 5 km |
| $c_{\mathrm{V}}$ | 75 m/s |
| $c_{\mathrm{A}}$ | 30 deg |
| $c_{\mathrm{D}}$ | 10 km |
| $D_{\mathrm{Lock}}$ | 11 km |
| $n_{\mathrm{t}}$ | 5 timesteps |

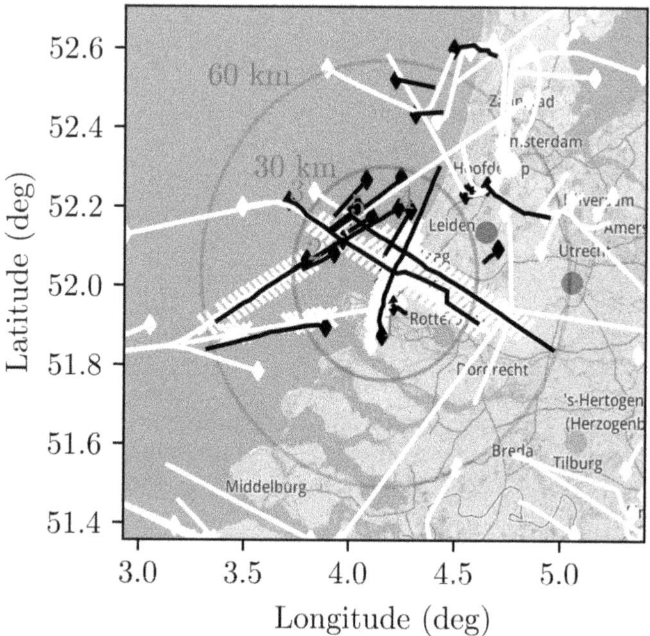


Figure 5.5.: Overview of all tracks found. The found tracks are marked in black and the reference tracks are marked in white. A diamond marks the starting point of each track.
Map Data ©OpenStreetMap contributors [92].

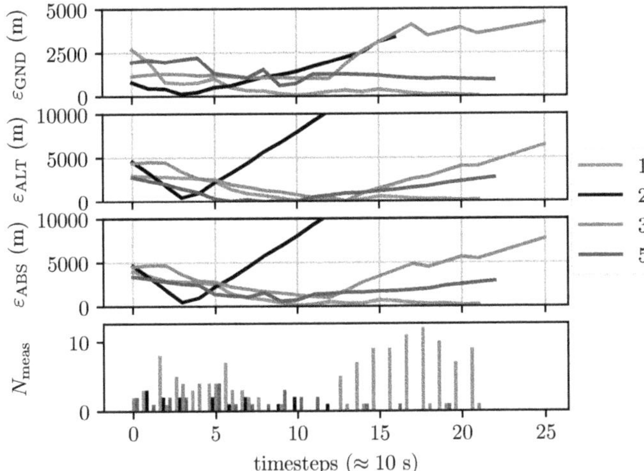


Figure 5.6.: Track errors for the 4 inspected tracks. The error in relation to the position above ground $\epsilon_{GND}$, the height above ground $\epsilon_{ALT}$, and the absolute error $\epsilon_{ABS}$ are shown. The bar graph at the bottom shows how many measurements could be assigned to the track at each time.

tracks. Thus it is possible to compare them with the recorded reference data[4]. The detailed comparison is shown in Figure 5.6. The top graph in this figure shows the position error in longitude and latitude ($\epsilon_{GND}$), the second graph shows the error in altitude ($\epsilon_{ALT}$) and the third graph shows the absolute position error ($\epsilon_{ABS}$). The bar graph shows how many range-Doppler measurements could be assigned to the track at each timestep[5].

For track 1, it can be seen in the figures 5.6 and 5.7 that the initialization of the track starts with an initial error of about 4000 m. Up to the eighth time step, the track is supplied with measurements and converges significantly. At time step nine, the target flies over the receiver position and thus enters the area shadowed by the antenna pattern. At this time, the position error $\epsilon_{GND}$ for this track is already smaller than 800 m. As a consequence of the shadowing, the track is not updated with measurements for the next 40 s. However, at this point the state of the Kalman filter is already sufficiently accurate that no increase in the position error can be observed.

After passing the receiver, the target enters the region predicted to be among the best detection performance in the modeling. In fact, the rate of measurements that can be assigned to the track increases significantly. At the best point, measurements from 12 of the 13 illuminators are assigned to the track. With these new measured values the absolute position error $\epsilon_{ABS}$ decreases to below 200 m. The comparison with the reference data then stops, as can be seen in Figure 5.6, because no further reference data is recorded. The

[4]Unfortunately, the OpenSky data did not contain elevation information for all relevant tracks. The examination is therefore based on the ADS-B reference data recorded by the system itself.

[5]Only one measured value can be assigned for each illuminator. Therefore, a maximum of thirteen measured values can theoretically be achieved.

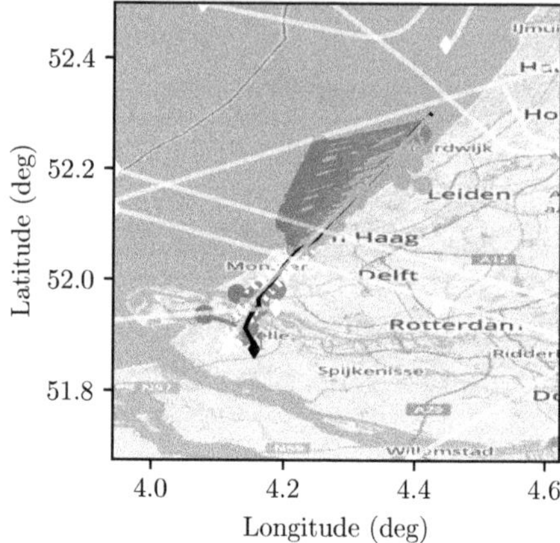


Figure 5.7.: Detailed view of track 1. The track is drawn in black. The colored points mark the estimated position of single measurements, the lines point to the position of the track to which they are associated.
Map Data ©OpenStreetMap contributors [92].

track continues as shown in Figure 5.5, well beyond the recorded reference data and stops at the end of the measurement interval.

Figure 5.7 also shows the error in the angle measurement[6] already suspected above. Systematic errors of different magnitude are found for different transmitter locations. The strongest deviation is found for the transmitter site "magenta". The estimated positions are clearly on a line that is approximately 15° counterclockwise from the track.
The three transmitters at location "red" produce an error in the same direction, but with estimated 5° it is much smaller. The results of the transmitter positions "orange" and "green" are much closer to the track, they also deviate predominantly counterclockwise, but also have measurements that deviate clockwise. Since multiple transmitters at different frequencies are considered at the "green" and "red" locations, the error does not appear to be frequency dependent.
Considering the transmitter locations in Figure 5.5, it appears that the error becomes larger with smaller angles between the LOS to the illuminator and the track.

The second discovered track 02, as seen in Figure 5.5, starts at the edge of the model's predicted operational area and then moves away in a southwestern direction, as shown in Figures 5.6 and 5.8. Therefore, only a few measurements can be assigned to the track. It starts with a position error of $\epsilon_{\mathrm{GND}} \approx 800\,\mathrm{m}$ and converges to $\epsilon_{\mathrm{GND}} \approx 200\,\mathrm{m}$ by the third time step. After that, however, the track starts to diverge again. It is noticeable that especially the error in the altitude information $\epsilon_{\mathrm{ALT}}$ increases to more than 10 km until the track stops, while the position error only increases to $\epsilon_{\mathrm{GND}} \approx 3.5\,\mathrm{km}$. The last update of

[6] An angular error shifts the result in the PA method on the bistatic ellipse of equal time difference of propagation, which is spanned with the transmitter and the receiver in the foci. The performance of the tracking algorithm is not affected by these angular errors, since the angular information is not used for this purpose.

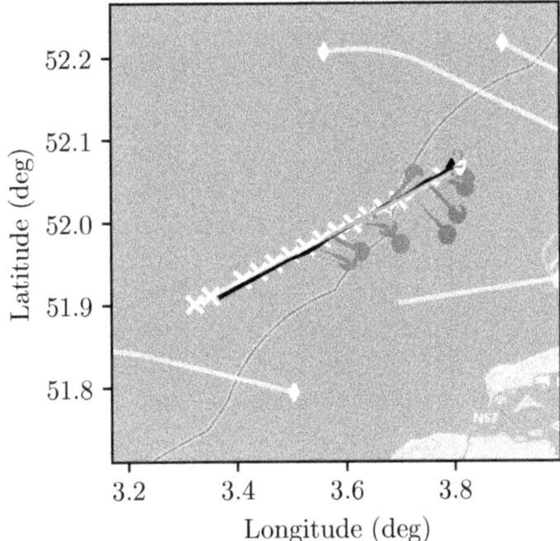


Figure 5.8.: Detailed view of track 2. The track is drawn in black. The colored points mark the estimated position of single measurements, the lines point to the position of the track to which they are associated.
Map Data ©OpenStreetMap contributors [92].

the track occurs at time step 12 and it terminates at time step 16.
The position estimates for this track are also mostly shifted counterclockwise from the actual position, as can be seen in Figure 5.8.

Track 3, as shown in Figure 5.5, starts northwest of the receiver and has a southeastern course passing over the receiver location. The track starts 5 time steps before the ADS-B reference could be received, as can be seen in the more detailed view in Figure 5.9.
Therefore, the comparison between the reference and the tracking data in Figure 5.6 already starts with a position error of only $\epsilon_{\text{GND}} \approx 1200\,\text{m}$. Up to the entry into the shadowed area above the receiver the track is updated regularly with several measurements. Thus it is converging further, which is especially noticeable in the altitude error. It drops to $\epsilon_{\text{ALT}} \approx 150\,\text{m}$ at timestep 12, whereas the position error remains relatively constant at $\epsilon_{\text{GND}} \approx 1000\,\text{m}$. The minimum absolute position error is $\epsilon_{\text{ABS}} \approx 1000\,\text{m}$.
After the overflight of the receiver station the track is updated only sporadically and also only with measurements from the illuminator at the location "magenta"[7]. As the track progresses, it diverges and terminates at time step 25 with an error $\epsilon_{\text{ABS}}$ greater than 5000 m.

Next consider track 5, which is parallel to track 3 and offset about 10 km to the north. Figure 5.10 shows this track in greater detail. This track also starts 4 time steps before the corresponding ADS-B reference data is available. Since this flight does not fly directly over the receiver, it is constantly supplied with new measurements in the first half of the measuring period.
It is noticeable that no measurements from the "red" position are used and that this track is also updated less frequently in the area east of the receiver. However, at that point the

[7] The track is now in the area southeast of the receiver for which a poor SNIR was modeled.

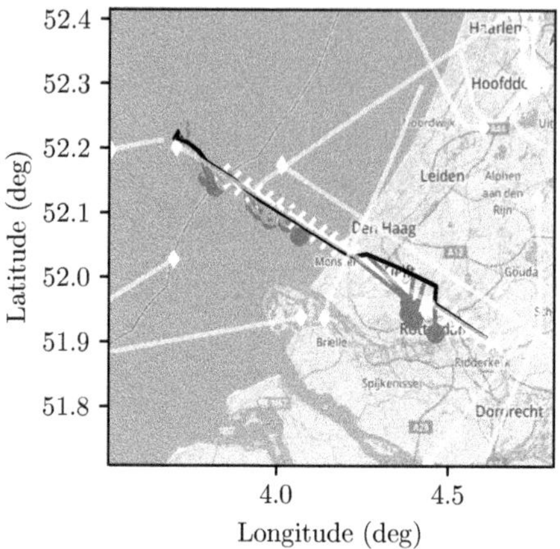


Figure 5.9.: Detailed view of track 3. The track is drawn in black. The colored points mark the estimated position of single measurements, the lines point to the position of the track to which they are associated.
Map Data ©OpenStreetMap contributors [92].

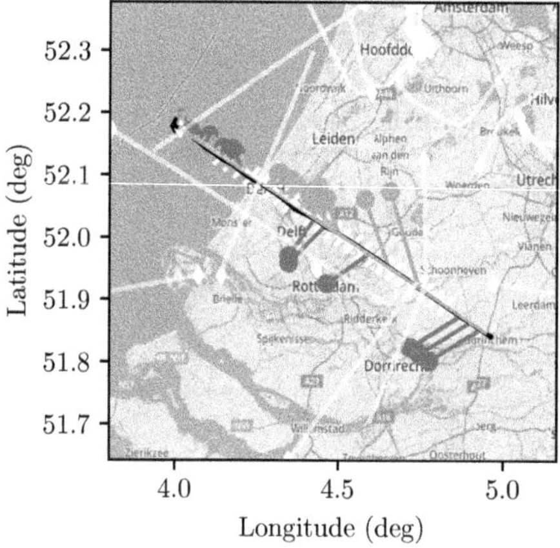


Figure 5.10.: Detailed view of track 5. The track is drawn in black. The colored points mark the estimated position of single measurements, the lines point to the position of the track to which they are associated.
Map Data ©OpenStreetMap contributors [92].

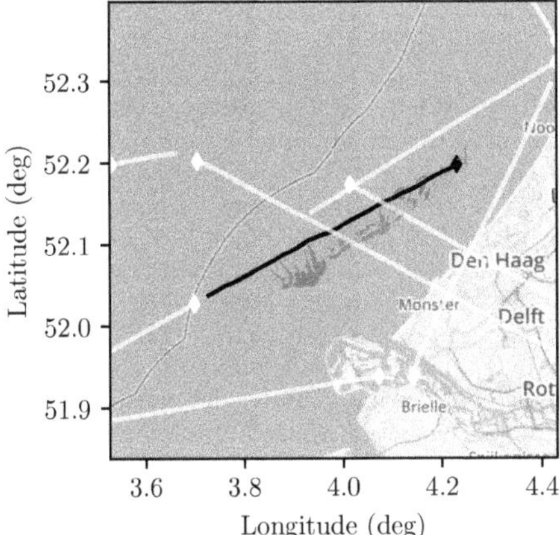


Figure 5.11.: Detailed view of track 4. The track is drawn in black. The colored points mark the estimated position of single measurements, the lines point to the position of the track to which they are associated.
Map Data ©OpenStreetMap contributors [92].

model of the UKF has converged sufficiently that the track does not diverge considerably. The reference measurement stops 4 time steps before the actual track. The track itself is kept alife further with three measurements from the location "magenta". It terminates only at the end of the measurement time.
The absolute position error of the track lies in the observed period between $700\,\mathrm{m} < \epsilon_{\mathrm{GND}} < 3.5\,\mathrm{km}$.

The last to be considered is track 4, which is presented in Figure 5.11. This track starts north of the receiver station and flies in a southwestern heading, on a trajectory similar to track 2. In the Figure 5.11 the track is plotted together with all reference data at the same time.
The startpoints of the reference tracks are marked with diamonds and all correspond to the same timestep as the origin of the track itself. It is thereby clear that this track does not correspond to any reference track in the OpenSky database or any recorded ADS-B measurement[8]. Therefore, no comparison with reference data is possible for this track. However, it is a nice example of the added value of independent airspace monitoring that does not rely on cooperative targets. The track starts shortly after the start of the measurement and is tracked until the end of the measurement time. It moves through the area of the measurement predicted as ideal SNIR by the model. The track is continuously updated with measurements from all illuminator locations. From the track data, the flight speed is $\approx 450\,\mathrm{km/h}$ and the flight altitude is between 5000 m and 6000 m.

[8]At the time of the measurements, there was no legal requirement to operate an ADS-B transponder.

## 5.4. Concluding Remarks

The results for the detections and tracking shown in the previous sections correspond well with the model described in Section 4.5. The predicted degradation of the performance in the direction to the transmitter is very apparent for all evaluated transmitter stations. The improvement predicted from the nulls to the sides of the array can also be seen in parts, but the detection range in these zones still seems to be modeled too optimistically. This indicates that the nulls of the real array are much less pronounced than in the modeling. All in all, the results show that it is beneficial to use as many illuminators as possible. The big advantage of the presented system is that the number of usable transmitters depends in principle only on the computing power of the system. The results also show that the system is currently not limited by the thermal noise power, but rather by the direct interference of the illuminator signal. This means, first of all, that the power of the transmitters has no direct influence on the detection range[9]. In addition, the results have shown that transmitters that are farther away actually tend to perform better than illuminators that are closer to the scenario[10]. This results in a large number of potentially usable illuminators, with which the surveillance area could be significantly increased. Since the number of usable transmitters scales directly with the computing power of the system, an improvement of the system performance could be realized comparatively inexpensive by this means.

The results of the tracking show considerable errors in the angle measurement. These cannot be explained by the large HPBW alone, since they occur too systematically. Also a constant offset, which could have arisen, e.g., by a rotation of the entire array, does not explain these errors, since such a constant error should affect all angle measurements equally. Also, there does not seem to be an influence of the transmitting frequency, since the angular error is at least similar for different transmitters from the same location. Instead, the error seems to depend on the position of the illuminator, or more precisely on the geometry between the transmitter, the target and the receiver. It could be an effect resulting from the parasitic signal power of the transmitter in the surveillance channel. This additional power could shift the angle during interpolation. So far, this effect has not been considered in detail and should be investigated in further studies.

Since the modeling described in Section 4.5 has been confirmed by the measurement results, recommendations for the improvement of the system performance can be derived from this model.
It has been shown that the separation of the surveillance channel and the reference channel is crucial for the system performance. To improve it, the first starting point is the receiving antenna array. By enlarging this array and, in the best case, also increasing the number of channels, the directivity could be increased. This would especially benefit the "blind" areas in the direction of the transmitter and reduce them significantly. In addition, it would be conceivable to exploit the polarization of the signals. This could be achieved by receiving the reference signals co-polarized and the target signals cross-polarized. To use this effect, different antennas (arrays) must be used for the surveillance and reference channels. This means, however, that the number of receivers would increase and thus also more computing power would be needed for the processing.
To minimize the number of receivers, an approach with arrays of different sizes for the surveillance- and reference- channels would be conceivable. However, the complexity of such a system would increase significantly in any case.

[9] This becomes relevant only at much larger ranges.

[10] In addition, but not yet taken into account in the model, distant transmitters may be shadowed by the curvature of the earth, while the illumination of higher flying targets remains LOS.

Since transmitters use different polarizations depending on the country or region, cross-polarized antennas would be a useful extension of the system, regardless of the use of polarization diversity.
In the context of polarization, another aspect that could be investigated is the behavior of the RCS as a function of polarization, so far only little literature can be found on this topic. Especially for the cross polarization of bistatic RCS there are hardly any sources available. With a better understanding or a better model for the RCS of different targets, the system model could be improved significantly.

The third important approach to improve the system performance lies in the method used for the digital beamforming. Here, more elaborate techniques, such as a linearly constrained minimum variance (LCMV) [24] beamformer could be employed. The aim of beamforming would be to shift a beampattern zero of the surveillance channels more specifically to the transmitter direction. As can be seen from the effect of the nulls of the current delay-and-sum beamformer in the model, the system performance could be improved significantly. This approach could be implemented with the least effort, and promises a significant gain for the range of the system.

The fourth important point in the processing chain is the adaptive filter for suppressing the reference signal in the surveillance channel. Currently, an NLMS algorithm is used. However, there are also various other approaches that could be used for this task [13, p.201] and potentially achieve a better suppression and thereby also a greater detection range.

The suspected influence of Doppler-affected clutter from the wind turbines in the north of the scenario could not be confirmed, since too little data is available for a detailed investigation. Nevertheless, a few observations could be made.
For one, the illuminators at the location "red" and "magenta" do indeed have a cluster of measurements in the area of the wind turbines. At the same time it can be seen in the tracking that the tracks started in this area cannot be assigned to any of the reference tracks.

The shown results in the area of system modeling and localization of the illuminators are the basis for the development of an actually independent, mobile passive detection system. Such a system could be placed at any location without prior knowledge of the scenario and without specialized expertise. It would autonomously search for available illuminators at the site, evaluate them, and optimize the detection area. Another logical extension for such a system would be the use of other illuminator systems like DAB or DVB-T. Together with a database of RCS models for different target types, such a self-configuring system could, after a short setup time, show the user the detection probabilities for the relevant targets.

# 6. Summary and Future Work

This thesis described the development of an experimental platform for passive detection techniques and a passive radar system for airspace surveillance.

In Chapter 3 the development of the passive detection system was described. The system is based on an 8-channel SDR synchronized for coherent reception. The synchronization procedure was shown in detail, and the implementation was verified with measurements. Alternative methods of synchronization have also been proposed in the appendix. For this core system, a modularly expandable and switchable antenna array was presented that allows fast switching between different modes or frequency ranges. The three antenna arrays that have been implemented so far were described. These are two ULAs, which are used for passive detection in the ISM bands at 1 GHz and 2.4 GHz. Additionally one UCA intended for passive detection and passive radar applications in the VHF band was included. To characterize the overall system, the noise figure in the selected bands was investigated, which yielded a sufficient value of 3.5 dB for the VHF band and 7.3 dB for the 2.4 GHz band.

Following the description and characterization of the hardware components, Chapter 4 described the signal processing of the passive radar system. The essential aspects were the preparation of the reference and surveillance signals and the calculation of the range-Doppler matrices as an intermediate result. The range-Doppler values were identified with a CA-CFAR for further processing. In order to determine the positions of the targets from this data, the position of the illuminators is required first.

Therefore, a new method is described by which the position of the illuminators can be determined from these measurements and ground-truth data of cooperative targets collected for this purpose. The PA method is used to approximate the position of the illuminators. Based on this approximation incorrect assignments in the data set are removed. With this reduced data set, a variety of solving methods for the determination of the position are compared. The technique was successfully demonstrated with a practical dataset achieving an accuracy better than 700 m. This procedure makes the system independent of external databases and thus facilitates deployment at new, unknown locations.

Once the positions of the illuminators were known, the targets can be tracked. For this purpose the range-Doppler measurements were clustered using the PA method. Each Cluster yielded an initial guess for position and velocity of the corresponding target which was used to initialize the Kalman based tracking algorithm. New measurements were then fused to the tracks improving the accuracy of the target model. With the results of real passive radar measurements, that were presented in Chapter 5 and the additionally collected ground truth information, it was proven, that for targets that were sufficiently updated, an absolute position accuracy better than 200 m was achieved.

Furthermore, a model for the system's performance was developed. In addition to the SNR, it also includes the influences of the direct interference of the illuminator signal into the surveillance channels, which is based on the antenna characteristics, the scenario geometry, and the signal processing. The developed model was compared to the real-world measurements and showed good agreement. Concluding, the results are discussed and the different constraints on the system are highlighted in order to give approaches for improvement. An important observation is that the system is currently not limited by thermal noise, but rather by the interference of the reference within the surveillance channel.

Even though the system more than met expectations in its first iteration, there is still some room for enhancement. First of all, there is still potential for improvement in the signal processing. For example, by using more complex beamforming techniques, the surveillance channels could be better separated from the reference signal. Likewise, in the suppression of the remaining reference signal in the surveillance signal, more advanced techniques could replace the NLMS algorithm. Considering the system performance model and comparing it to the limit given by thermal noise, it is apparent that an improvement in range could be achieved with these measures. In addition to the enhancements in the signal processing, it has also been shown that the location, or rather the distance, of the transmitters also have an influence on the performance of the system. This relationship is already well represented by the model created. Furthermore, the influence of the effective bandwidth on the detection performance was observed in the measurements. By combining these properties with the presented method of illuminator localization, a completely autonomous illuminator selection system could be created for a next iteration of the system.

Another promising field for future research is the further use of ground truth data. In addition to the localization of the illuminator positions shown, it would also be conceivable to monitor or even calibrate other system parameters using this data set. For example, antenna patterns could be monitored, which would allow the beamformer settings to be adjusted at runtime if necessary. Additionally, a control loop could also be designed to provide better suppression of the reference signal in the surveillance signal during beamforming.
There are also still some opportunities for further development in the area of tracking. For example, the as yet unused angular information could be evaluated as a further measured parameter. In this context, the extension of the antenna array and the optimization of its modeling could be investigated and effects such as the mutual coupling of elements investigated in more detail.

# Appendices

# A. 14 GHz FMCW Radar

In preparation for the work on the passive radar system presented in this thesis, initial findings were collected using an FMCW radar. Therefore, it is presented here briefly. The active FMCW radar system was intended for use as a collision warning radar in small aircraft. Collisions with other aircraft occur mainly in the lower airspace E, where there is a mixture of flights under visual flight rules (VFR) and flights under instrument flight rules (IFR) monitored by air traffic control (ATC). The systems available so far are cooperative systems that are not able to actively detect obstacles but rely on information sent by other aircraft for their function. On the one hand, there is the airborne collision avoidance system (ACAS), which uses the position reports of secondary radar transponders or ADS-B [95]. There is no obligation for such equipment in the lower airspace. On the other hand, there is the FLARM system, which provides a similar functionality. This system was originally developed for gliding to minimize the risk of collision when climbing together in narrow updrafts but quickly became widespread throughout the general aviation sector. Compared to these systems, an active radar has the advantage of being able to detect passive obstacles such as mountains or television towers or aircraft that are not equipped with one of the above-mentioned systems.

The system is designed as an FMCW radar (see Chapter 2.3) and has the possibility to use digital beamsteering on the receive side. The developed system was extended within the scope of student theses [96, 97]. In the context of this work, the system is only an intermediate step in the development of the passive radar system. Nevertheless, a short overview of the developed system and a small insight into the results shall be given here. A block diagram of the developed prototype is shown in Figure A.1.
To keep the system cost effective and in preparation for future expansion, it is equipped with 8 parallel power amplifiers, each providing one watt of transmit power. On the receiver side, four parallel channels are provided, which are evaluated by a computer and thus enable digital beamforming applying the method described in Section 2.6.

All signal generation and functions of the radar are configured and controlled by an Atmel 8-bit microcontroller (MEGA128). The signal source is based on an Analog Devices HMC531LP5 voltage-controlled oscillator (VCO), which allows a frequency range between 13.6 GHz and 14.9 GHz. The VCO is controlled by an Analog Devices ADF4159 PLL, which uses the 1/4 clock output of the VCO for its feedback loop. This PLL can be configured to directly generate the frequency sweeped radar signal.
As a reference, the Vectron temperature compensated crystal oscillator (TCXO) TX-500 with 100 MHz is used, since it has particularly low phase noise. The generated signal is split to the transmit and receive branches by several Wilkinson splitters. The resulting losses are compensated by a series of Mini-Circuits pre-amps (AVA-183A+). For the eight one watt power amplifiers, Qorvo TGA2527-SM amplifiers are used.

On the receiving path, the antenna signal is initially amplified with an Analog Devices HMC903LP3E as low noise amplifier (LNA). This is followed by a bandpass filter in microstrip technology designed in Keysight ADS (see Figure A.2), which is used to filter

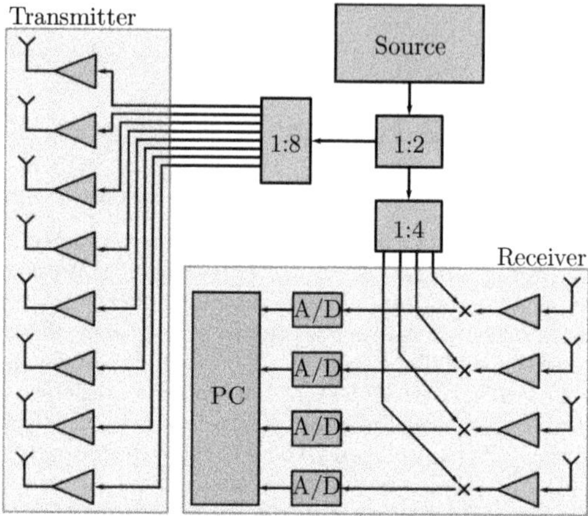


Figure A.1.: Block diagram of the FMCW radar

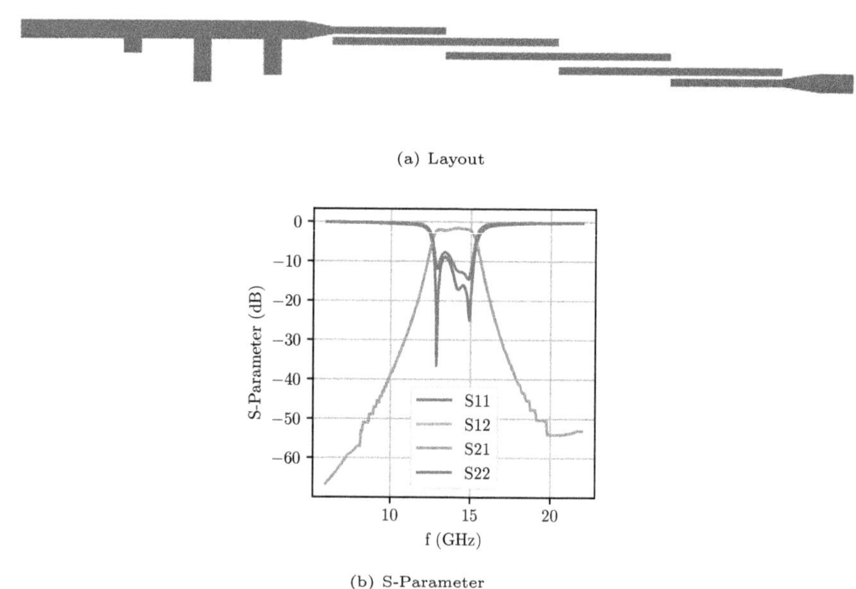


(a) Layout

(b) S-Parameter

Figure A.2.: Layout and S-Parameters of the developed 14 GHz Bandpass.

Figure A.3.: Picture of the FMCW radar PCB

interfering signals in adjacent bands. The power of the filtered signal is then measured using an integrated directional coupler and power measurement IC (Macom MACP-010563), this measurement is the basis for an AGC in the baseband amplifiers. The signal is then mixed into the complex I and Q components of the baseband using an Analog Devices HMC1113LP5E complex mixer. The baseband processing consists of a cascade of four Analog Devices ADA4841 operational amplifiers that amplify the signal and compensate for the distance dependency, by applying a high-pass characteristic. Details may be found in the Master thesis by F. Schwartau [98]. Last in the chain is a Texas Instruments PGA112 programmable amplifier, which is used together with the power detector mentioned above to control the output signal to match the input voltage range of the following A/D converters. In the system, a Saleae Logic Pro 16 logic analyzer is used as a parallel A/D converter. A picture of the developed board is depicted in Figure A.3.

For the operation of the system it is important to know the distance at which targets of a certain size can still be detected. To find out, the radar Equation (2.16) can be used.
It is assumed that an SNR of 10 dB is required for detection and the antennas have a gain of $G_{\mathrm{I}} = G_{\mathrm{R}} = 15\,\mathrm{dB}$. For the influence of noise, we assume a noise figure of $F = 3\,\mathrm{dB}$ and a noise temperature of $T_0 = 290\,\mathrm{K}$. The operating frequency of the system is chosen to be 14 GHz. With this data, the achievable range can be plotted as a function of RCS. The results are shown in Figure A.5. The curve for a transmit power of 8 W which corresponds to the finished prototype is shown in orange. Also shown are the curves for 1 W, which corresponds to operation with a single transmit channel, and as an additional reference the curve that would result with a transmit power of 100 W.

Parallel to the hardware development, work is also carried out on the signal processing. For this purpose, measurements were made with an early prototype of the system. This

Figure A.4.: Measurement setup with the prototype

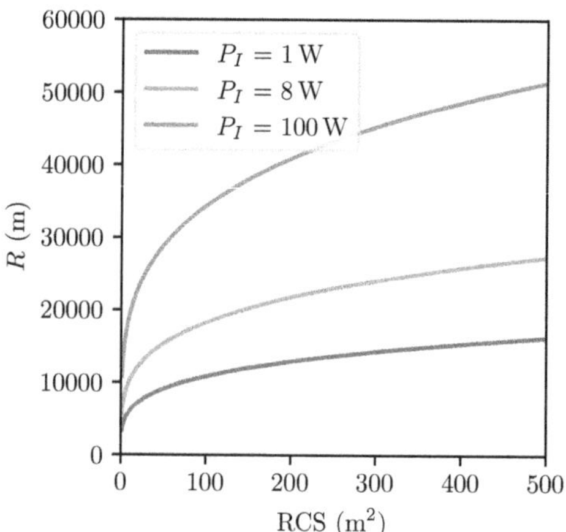


Figure A.5.: Achievable Range over the RCS for the FMCW radar system.

prototype included only one channel each for the transmitter and receiver. The antenna system comprised two horn antennas with approximately 15 dB gain (see Figure A.4). These were manually tracked to the target.

The measurements were performed at the Porta Westfalica airfield, Germany, where a motor glider of the type Scheibe Falke 25C (see Figure A.4) was used as a single target. An exemplary set of the performed measurements is shown in Figure A.6. In Figure A.6(a) the position of the radar system is plotted as a black dot. The blue dots show the ground truth position of the target, which was recorded with a GPS logger. In Figure A.6(c) and A.6(d) the radargrams for the upsweep and downsweep are shown. In each of these, the tracking of the antennas can be seen by the change in the static background clutter. On the basis of this data, a data processing system was developed within the scope of the diploma thesis of Inna Frede [96].

The data processing achieves clutter suppression by means of an ordered-statistic constant false-alarm rate (OS-CFAR), associates the targets from the two sweeps and implements a Kalman filter for 1-dimensional tracking of the range. This Kalman filter works similar to the example described in Section 2.8. The results can be seen in Figure A.6(b). The purple dots represent the range measurements extracted from the radargrams by the OS-CFAR algorithm. The blue dots show the distance of the target, calculated from the ground truth. The black line shows the result of the Kalman filter.

These first practical tests already show the importance of signal processing for clutter suppression and target extraction, in the evaluation of radar data. As can be seen in the Figures A.6(c) and A.6(d), the target signal is strongly masked by the clutter. The CFAR algorithm, as shown in Figure A.6(b), already provides a much more precise result, which in turn could be further improved by the use of the Kalman filter. This is still a very simple model of Kalman filtering, which could be further improved e.g. by extending it to a complete three-dimensional model of the aircraft position or by adding further information, such as the angle. A tracking of several aircraft positions in three-dimensional space is shown in Chapter 4.4 for a passive radar system.

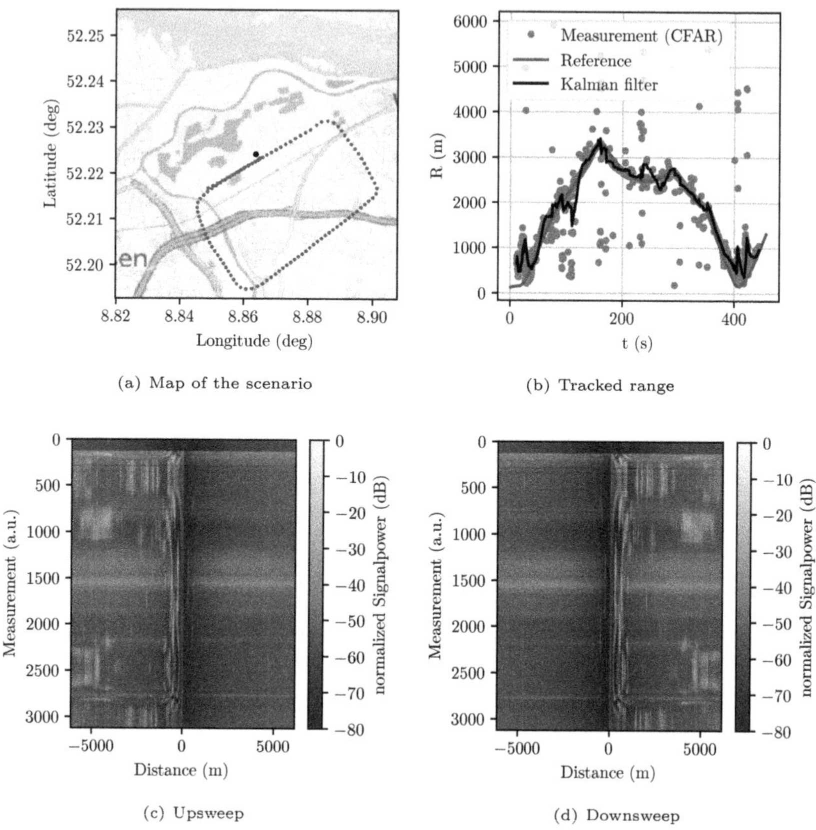


(a) Map of the scenario

(b) Tracked range

(c) Upsweep

(d) Downsweep

Figure A.6.: Evaluation of measurement data from the FMCW radar. The measurement data for up- and down-sweep plotted over time, the positions of the target during the measurement as well as the results or intermediate steps of the signal processing are shown. Map data ©OpenStreetMap contributors [92].

In the context of another master thesis by Marvin Wenzel [97], the existing system was extended with a beam steering for the transmit channel by means of a Rotmann lens. The developed lens can be seen in Figure A.7. The radiation characteristic measured for the combination of the Rotman lens and the used array of patch antennas can be seen in Figure A.8. The eight distinct main lobes that lie between ±45° are clearly visible here. Finally, in figure A.9 the result of a measurement on the open field is shown, with which the two-dimensional beam steering of the system was verified.

Figure A.7.: Picture of the Rotman Lens [97].

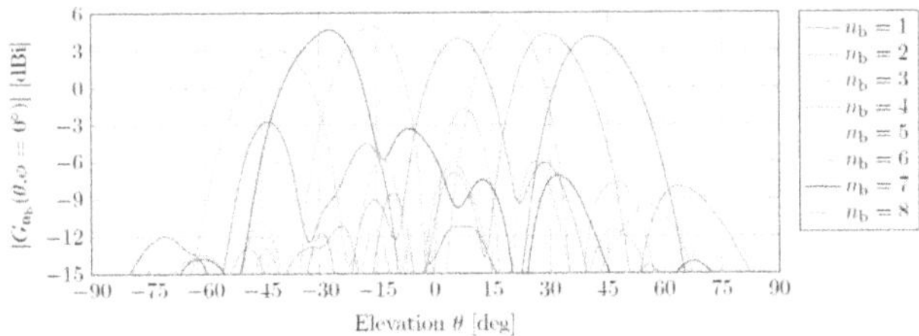



Figure A.8.: The measured radiation characteristic of the combination of the radiation characteristic of Rotman lens and the used array of patch antennas. [97]

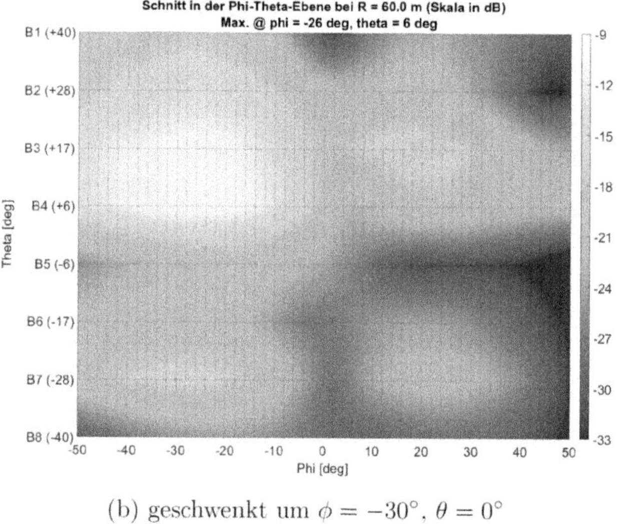


(b) geschwenkt um $\phi = -30^\circ$, $\theta = 0^\circ$

Figure A.9.: Exemplary result of an open field measurement with the 2D scanned array. The target was located at a distance of 60 m [97].

# B. Synchronization of the SDR System using the internal PLLs

## B.1. LO generation using the internal PLLs

If the daughterboard's internal PLLs are used, the LO signals are derived from a common reference signal ($\text{Ref}_{\text{in}}$ in Figure 3.2). As a consequence the LOs' phases are locked among each other. Regarding synchronization, the problem with this method is that a locked phase does not necessarily mean that there is no phase ambiguity between the reference signal and the LO signal, which would result in a phase ambiguity between the system's channels.[1] To avoid these phase ambiguities, it would be possible to use an additional synchronization signal or to restrict the divider ratios to certain special cases, where no additional synchronization is necessary (cf. [99]).

To elaborate this point, consider the well-known block diagram of a standard PLL circuit depicted in Figure B.1.[2] Two dividers, namely the reference divider with a divider ratio $R$ and the feedback divider with a divider ratio of $N$ realize the relation

$$f_{LO} = \frac{N}{R} \cdot f_{ref} \tag{B.1}$$

between the reference signal and the LO signal.[3]

If the PLL is locked, the phase relation

$$\frac{1}{N}\varphi_{LO} + \frac{2\pi}{N}c_1 = \frac{1}{R}\varphi_{ref} + \frac{2\pi}{R}c_2, \tag{B.2}$$

applies at the phase detector where $2\pi/N \cdot c_1$ and $2\pi/R \cdot c_2$ represent phase ambiguities introduced by the frequency dividers. The coefficients $0 \leq c_1 \leq N-1$ and $0 \leq c_2 \leq R-1$ are integers which are arbitrarily determined during the PLL's startup.

Rearranging (B.2) for $\varphi_{LO}$ yields

$$\varphi_{LO} = \frac{N}{R}\varphi_{ref} + 2\pi\left[\frac{N}{R}c_2 - c_1\right] \tag{B.3}$$

and by defining

$$O = 2\pi\left[\frac{N}{R}c_2 - c_1\right] \tag{B.4}$$

[1] For completeness, it should be mentioned that such a phase ambiguity could be corrected by performing a calibration of the system after each frequency change, as is demonstrated in [42].

[2] Only integer-N PLLs are considered here.

[3] Both dividers are integer values.

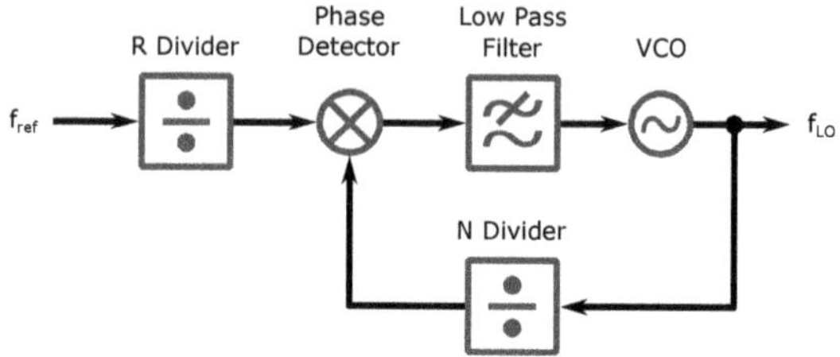


Figure B.1.: Block diagram of an integer-N PLL, after [100]

for the additional arbitrary phase offset, (B.3) is rewritten as

$$\varphi_{LO} = \frac{N}{R}\varphi_{ref} + O. \tag{B.5}$$

For the given applications, a well-defined phase relation between $\varphi_{ref}$ and $\varphi_{LO}$ is mandatory and thus

$$O = 2\pi\left[\frac{N}{R}c_2 - c_1\right] \overset{!}{=} 2\pi k \tag{B.6}$$

is required, where $k$ is an arbitrary integer. From (B.6) follows directly that

$$k = \left[\frac{N}{R}c_2 - c_1\right] \tag{B.7}$$

applies, which may only be an integer value if the ratio $N/R$ is integer.

Several important conclusions can be drawn from this finding: If the reference divider is disabled ($R = 1$), a well-defined phase relation is guaranteed for every $N$. In contrast, if both the reference and feedback divider are used, only integer ratios $N/R$ ensure the desired phase relation.

For the general case of arbitrary (i.e. possibly non-integer) ratios $N/R$ the number of the PLL's ambiguous locking states can be determined by careful analysis of (B.4).

By rearranging (B.4) as

$$O = 2\pi \cdot \frac{N}{R}c_2 - 2\pi \cdot c_1 \tag{B.8}$$

it can be seen that only the arbitrary coefficient $c_2$ introduced by the reference divider can cause a phase ambiguity which is not an integer multiple of $2\pi$, while the coefficient $c_1$ introduced by the feedback divider leads to a negligible ambiguity of integer multiples of $2\pi$.

Thus the non-negligible phase offset is defined as

$$O' = 2\pi \cdot \frac{N}{R} \cdot c_2. \tag{B.9}$$

From (B.9) it is obvious that the number of possible ambiguous states is equal to [4]

$$M = \frac{R}{\gcd(N, R)} \tag{B.10}$$

[4] gcd is the greatest common divisor.

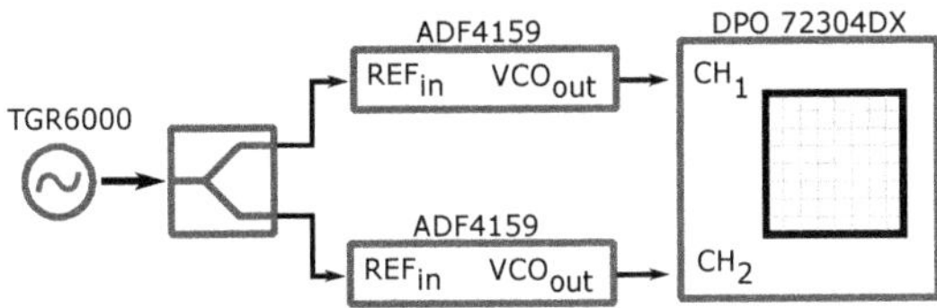


Figure B.2.: Block diagram of the phase ambiguity measurement setup

which is in fact the denominator of the reduced fraction $N/R$. While using PLLs with $N$ and $R$ being chosen correctly ensures both coherence and ambiguity-free phase relations between the individual channels, the limitations regarding proper values for $N$ and $R$ drastically decrease the system's flexibility because the output frequency is limited to integer multiples of the reference frequency as can be seen from (B.1).

At first sight, using fractional-N PLLs rather than integer-N PLLs might alleviate those limitations to some extend. However, because the internal state of the fractional-N feedback dividers can usually not be synchronized [101] a well-defined phase relation is not guaranteed.

## B.2. Phase ambiguities of integer-N PLLs

For this measurement, two ADF-4159 evaluation boards are used to achieve a synchronized setup. Both boards are driven with a common 100 MHz reference frequency, generated with a TTI TGR6000 signal source. To evaluate the phase relationship between the two signals, the VCO outputs are recorded with a Tektronix DPO72304DX oscilloscope. The measurement setup is depicted in Figure B.2. For each data point, the PLLs were forced to re-lock by toggling the reference signal. After the PLLs had sufficent time to lock again, the phase difference between the two VCO signals is examined. Each measurement consists of 2000 data points, where the phase difference of each data point relative to the first measurement is evaluated. In Figure B.3 the probability of the phase difference is shown for different settings of the $R$ and $N$ divider. It is clear, that if the restrictions described in Section 3.1.2 are met (measurement A and B) there is no phase ambiguity. For the divider combination R=2 and N=241 (measurement C), two phase states are possible ($0^\circ$ and $180^\circ$), as it is described in equations B.9 and B.10. The same is true for measurement D, where four phase states ($0^\circ$, $90^\circ$, $180^\circ$ and $270^\circ$) can be observed for a divider combination of R=4 and N=121.

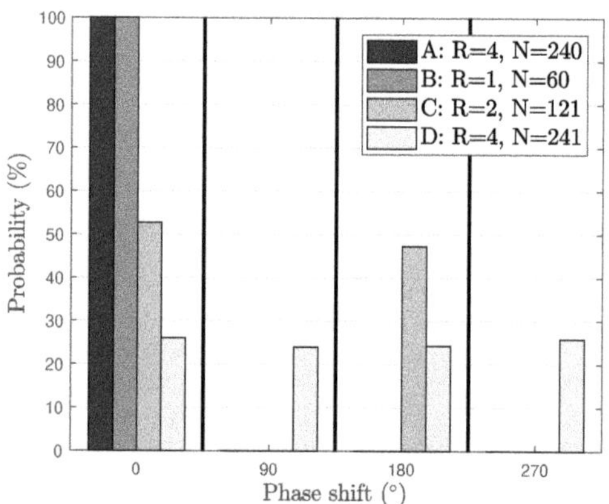


Figure B.3.: Distribution of the phase error of two integer-N PLLs in steps of 90 ° for different divider values

# C. Frontend Measurements

The gain curve of the front-ends described in Chapter 3.1.3 is measured for the two realized variants. The results of this measurement are shown in graph C.1.

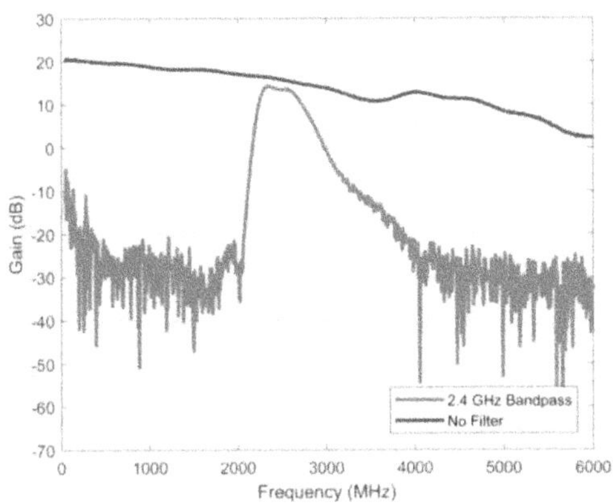


Figure C.1.: Plot of the gain of the analog front ends.

## C.1. Array Gain Plots

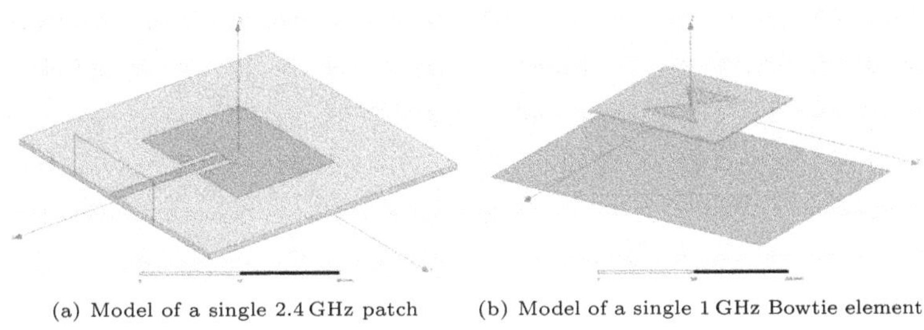

(a) Model of a single 2.4 GHz patch

(b) Model of a single 1 GHz Bowtie element

Figure C.2.: Characteristics of the modular linear arrays

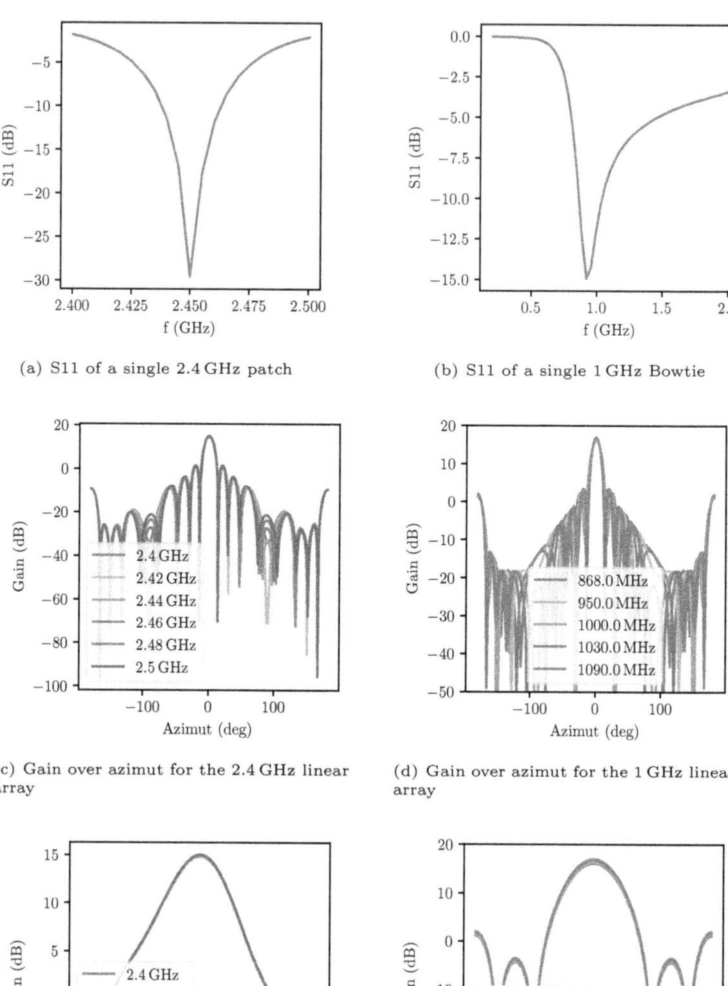


(a) S11 of a single 2.4 GHz patch

(b) S11 of a single 1 GHz Bowtie

(c) Gain over azimut for the 2.4 GHz linear array

(d) Gain over azimut for the 1 GHz linear array

(e) Gain over elevation for the 2.4 GHz linear array

(f) Gain over elevation for the 1 GHz linear array

Figure C.3.: Characteristics of the modular linear arrays

# D. Passive detection

Passive detection means determining the position of a target without actively illuminating it as in a classic radar. Some targets are "cooperative" and actively transmit a signal themselves. In the best case, these signals contain the position data, as is the case with ADS-B transmitters used in aviation. However, less cooperative targets, such as the UAV class, which has become increasingly popular in recent years, also frequently emit signals. These are mostly telemetry data or broadband transmissions of high-resolution video streams. Especially for this class, passive detection is of interest since these small targets are difficult to distinguish from clutter due to their extremely small radar cross-section and their often low velocities and thus only small Doppler shifts.

The hardware as introduced in Chapter 3 has all elements needed for such a detection system: sensitive receivers and an array of antennas to allow estimation of incident angle. The passive detection is based on the beamforming described in Section 2.6 and applies it to the eight coherently received channels. With a single sensor, using a single linear array only measurements of the bearing towards the target are possible, but neither distance, speed, or elevation can be measured. Although, if the transmit power of the target is known, a statement about the distance could also be made. But since the antenna characteristics of the target and the exact orientation must also be assumed to be unknown, such an estimation is only possible within very wide ranges. In the context of this work, passive detection is therefore only a first step on the way to the development of a passive radar system and serves as proof that the coherent receiver works correctly[1].

An example of the application of passive detection is shown in Figure D.1. Figure D.1(a) shows a graphical user interface (GUI) that is used to visualize the measurements. On the top left is the monitored spectrum; in this case, the 2.4 GHz ISM band can be seen. Conjoined frequency regions, which are above the average noise level, are combined to a single target, which is indicated by the blue sections in the spectrum. In the lower-left corner, the current spectrum is then shown over angle. Here the measurement of a wireless local area network (W-LAN) channel in the building in direct LOS of the system can be seen. In addition, the wide video downlink of a UAV from the manufacturer DJI can be seen. The measurements are also processed and plotted as lines of bearing on the map shown to the right. The color of the line encodes the frequency, and old measurements have an "afterglow" and fade after some time. In the Figures D.1(b) and D.1(c) the ground truth data of the target is added retrospectively. In D.1(b) it can be observed that the target can be detected well if there is LOS propagation. As soon as the target is out of the LOS area and disappears behind a building, the reflections of the signal from the buildings behind it dominate, as can be seen in D.1(c). In this case, a clear bearing is no longer possible and ghost targets occur. Despite its drawbacks, the method is well suited for the detection of UAVs. Especially when several sensors are involved and thus methods like cross bearing are possible, very good detection results can be achieved at long ranges.

[1] One approach to actually using this technique would be to deploy a distributed system with multiple sensors; such an approach to UAV detection and deterrence is being explored, for instance, by projects such as "Abwehr von unbemannten Flugobjekten für BOS (AMBOS)" [102].

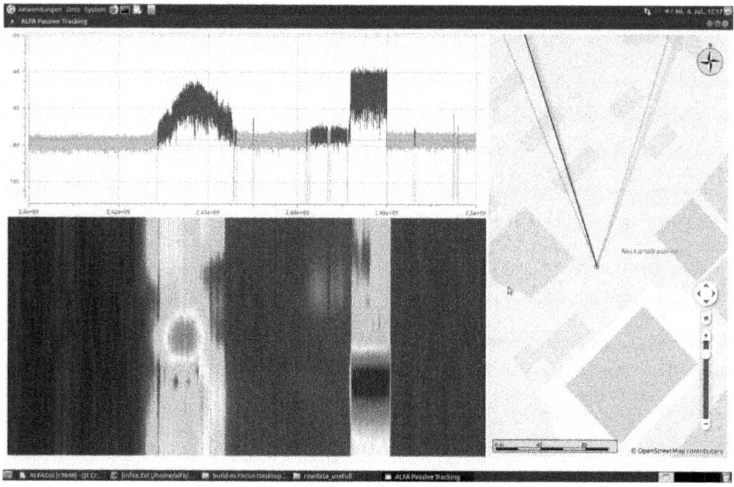

(a) Screenshot of the GUI during a demonstration.

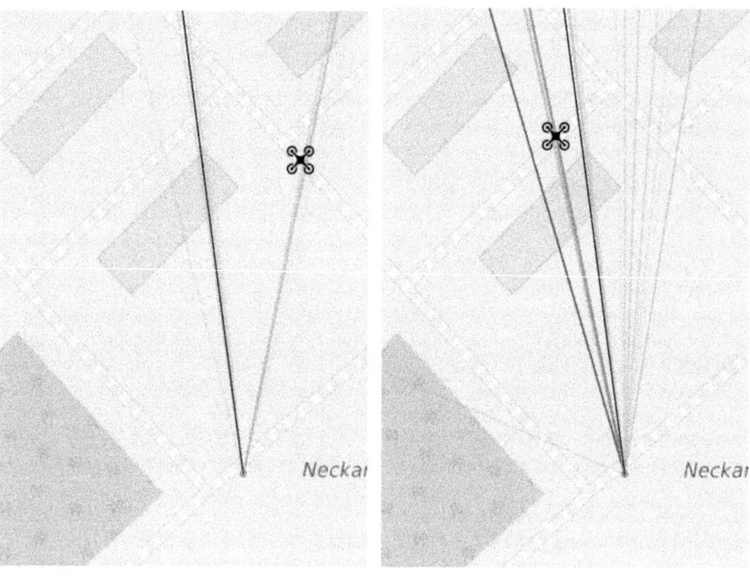

(b) Result with LOS to target

(c) Result with Multipath to target

Figure D.1.: Exemplary Measurements with the passive Detection Map Data ©OpenStreetMap contributors [92].

# List of References


[1] M. Skolnik, *Radar Handbook*, 3rd ed. Mc Graw Hill, 2008.

[2] Hertz, *Die elektrischen Wellen (1888).* Berlin, Heidelberg: Springer Berlin Heidelberg, 1953, pp. 169–178. [Online]. Available: https://doi.org/10.1007/978-3-642-86911-2_27

[3] H. Griffiths, P. Knott, and W. Koch, "Christian Hülsmeyer: Invention and Demonstration of Radar, 1904," *IEEE Aerospace and Electronic Systems Magazine*, vol. 34, no. 9, pp. 56–60, 2019.

[4] H. Griffiths, "Early history of bistatic radar," in *2016 European Radar Conference (EuRAD)*, 2016, pp. 253–257.

[5] H. Griffiths, "Multistatic, MIMO and networked radar: The future of radar sensors?" in *The 7th European Radar Conference*, 2010, pp. 81–84.

[6] W. Wiesbeck and L. Sit, "Radar 2020: The future of radar systems," in *2014 International Radar Conference*, 2014, pp. 1–6.

[7] "MSSR M10R," https://www.ramet.as/mssr-m10sr, accessed: 2022-05-30.

[8] H. Kuschel and D. O'Hagan, "Passive radar from history to future," in *11-th INTERNATIONAL RADAR SYMPOSIUM*, 2010, pp. 1–4.

[9] H. D. Griffiths and C. J. Baker, *An introduction to passive radar.* Artech House, 2017.

[10] M. Malanowski, K. Kulpa, and J. Misiurewicz, "PaRaDe - PAssive RAdar DEmonstrator family development at Warsaw University of Technology," in *2008 Microwaves, Radar and Remote Sensing Symposium*, 2008, pp. 75–78.

[11] P. Knott, T. Nowicki, and H. Kuschel, "Design of a disc-cone antenna for Passive radar in the DVB-T frequency range," in *2011 German Microwave Conference*, 2011, pp. 1–4.

[12] A. Capria, D. Petri, C. Moscardini, M. Conti, A. C. Forti, R. Massini, M. Cerretelli, S. Ledda, V. Tesi, E. Dalle Mese, G. B. Gentili, F. Berizzi, M. Martorella, R. Soleti, T. Martini, and A. Manco, "Software-defined Multiband Array Passive Radar (SMARP) demonstrator: A test and evaluation perspective," in *OCEANS 2015 - Genova*, 2015, pp. 1–6.

[13] M. Malanowski, *Signal Processing for Passive Bistatic Radar.* Artech House, 2019.

[14] "Advanced Low Flying Aircrafts Detection and Tracking," https://alfa-h2020.eu/, accessed: 2021-04-26.

[15] M. I. Skolnik, *Introduction to Radar Systems.* McGraw-Hill Book Company, Inc., 1981.

[16] R. Burkholder, L. Gupta, and J. Johnson, "Comparison of monostatic and bistatic radar images," *IEEE Antennas and Propagation Magazine*, vol. 45, no. 3, pp. 41–50, 2003.

[17] M. Jankiraman, *Design of multi-frequency CW radars.* SciTech Publishing, 2007, vol. 2.

[18] D. G. Mixon, "Doppler-only multistatic radar," 2006.

[19] M. Jackson, "The geometry of bistatic radar systems," in *IEE proceedings F (communications, radar and signal processing)*, vol. 133, no. 7. IET, 1986, pp. 604–612.

[20] J. Ville, "Theory and application of the notion of complex signal," RAND CORP SANTA MONICA CA, Tech. Rep., 1958.

[21] P. Woodward, *Probability and Information Theory, with Applications to Radar.* Elsevier, 1964.

[22] L. Spafford, "Optimum radar signal processing in clutter," *IEEE Transactions on Information Theory*, vol. 14, no. 5, pp. 734–743, 1968.

[23] J.-R. Ohm and H. D. Lüke, *Signalübertragung: Grundlagen der digitalen und analogen Nachrichtenübertragungssysteme.* Springer-Verlag, 2015.

[24] H. Van Trees, *Optimum Array Processing: Part IV*, ser. Detection, Estimation, and Modulation Theory. Wiley, 2004.

[25] R. J. Mailloux, *Phased array antenna handbook.* Artech house, 2018.

[26] C. A. Balanis, *Antenna Theory: Analysis and Design*, 4th ed. Wiley, 2015.

[27] F. Schwartau, "Large Aperture Array Radar Systems for Automotive Applications," dissertation, TU Braunschweig, 2021.

[28] R. E. Kalman, "A new approach to linear filtering and prediction problems," *Journal of Basic Engineering*, 1960.

[29] R. Marchthaler and S. Dingler, *Kalman-Filter.* Springer, 2017.

[30] R. Labbe, "Kalman and bayesian filters in python," *Chap*, vol. 7, p. 246, 2014.

[31] M. S. Grewal and A. P. Andrews, *Kalman filtering: Theory and Practice with MATLAB.* John Wiley & Sons, 2014.

[32] A. C. Rencher and G. B. Schaalje, *Linear models in statistics.* John Wiley & Sons, 2008.

[33] R. Van Der Merwe, "Sigma-point Kalman filters for probabilistic inference in dynamic state-space models," Ph.D. dissertation, OGI School of Science & Engineering at OHSU, 2004.

[34] M. I. Ribeiro, "Kalman and extended kalman filters: Concept, derivation and properties," *Institute for Systems and Robotics*, vol. 43, 2004.

[35] J. K. Uhlmann, "Dynamic map building and localization: New theoretical foundations," Ph.D. dissertation, University of Oxford Oxford, 1995.

[36] S. J. Julier and J. K. Uhlmann, "New extension of the Kalman filter to nonlinear systems," in *Signal processing, sensor fusion, and target recognition VI*, vol. 3068. International Society for Optics and Photonics, 1997, pp. 182–193.

[37] S. J. Julier, "The scaled unscented transformation," in *Proceedings of the 2002 American Control Conference (IEEE Cat. No. CH37301)*, vol. 6. IEEE, 2002, pp. 4555–4559.

[38] M. Krueckemeier, F. Schwartau, C. Monka-Ewe, and J. Schoebel, "Synchronization of Multiple USRP SDRs for Coherent Receiver Applications," in *2019 Sixth International Conference on Software Defined Systems (SDS)*. IEEE, 2019, pp. 11–16.

[39] H. Fu, S. Abeywickrama, L. Zhang, and C. Yuen, "Low-Complexity Portable Passive Drone Surveillance via SDR-Based Signal Processing," *IEEE Communications Magazine*, vol. 56, no. 4, pp. 112–118, April 2018.

[40] and J. Lu, X. Tian, G. Chen, K. Pham, and E. Blasch, "Cognitive radio testbed for Digital Beamforming of satellite communication," in *2017 Cognitive Communications for Aerospace Applications Workshop (CCAA)*, June 2017, pp. 1–5.

[41] M. Hua, C. Hsu, W. Liao, C. Yao, T. Yeh, and H. Liu, "Direction-of-Arrival Estimator using Array Switching on Software Defined Radio Platform," in *2011 IEEE International Symposium on Antennas and Propagation (APSURSI)*, July 2011, pp. 2821–2824.

[42] M.-C. Hua, C.-H. Hsu, and H.-C. Liu, "Implementation of Direction-of-Arrival Estimator on Software Defined Radio Platform," in *2012 8th International Symposium on Communication Systems, Networks Digital Signal Processing (CSNDSP)*, July 2012, pp. 1–4.

[43] E. Research, *TwinRX IF Board Rev C*, 2017. [Online]. Available: https://files.ettus.com/schematics/twinrx/TwinRXIFBoardRevC.pdf

[44] E. Research, *TwinRX RF Board Rev D*, 2017. [Online]. Available: https://files.ettus.com/schematics/twinrx/TwinRXRFBoardRevD.pdf

[45] N. Temple, "Phase Synchronization Capability of TwinRX Daughterboards and DoA Estimation," Ettus Research, Tech. Rep., 2016. [Online]. Available: https://github.com/EttusResearch/gr-doa/blob/master/docs/whitepaper/doa_whitepaper.pdf

[46] H. Tang, Z. Nie, and X. Zong, "Approximate Calculation of HPBW for Uniform Circular Array," in *2018 Cross Strait Quad-Regional Radio Science and Wireless Technology Conference (CSQRWC)*, 2018, pp. 1–3.

[47] H.-G. Unger, *Hochfrequenztechnik in Funk und Radar*. Springer, 1988.

[48] K. Loi, S. Uysal, and M. Leong, "Design of a wideband microstrip bowtie patch antenna," *IEE Proceedings-Microwaves, Antennas and Propagation*, vol. 145, no. 2, pp. 137–140, 1998.

[49] M. Pozar, "Microwave engineering 4th edn, Ch. 2, 51–56," 2009.

[50] K. Technologies, "Noise Figure Measurement Accuracy: The Y-Factor Method," Keysight Technologies, Tech. Rep., 2020. [Online]. Available: https://www.keysight.com/de/de/assets/7018-06829/application-notes/5952-3706.pdf

[51] H. Griffiths and N. Willis, "Klein Heidelberg—The First Modern Bistatic Radar System," *IEEE Transactions on Aerospace and Electronic Systems*, vol. 46, no. 4, pp. 1571–1588, 2010.

[52] F. Pieralice, F. Santi, D. Pastina, M. Bucciarelli, H. Ma, M. Antoniou, and M. Cherniakov, "GNSS-based passive radar for maritime surveillance: Long integration time MTI technique," in *2017 IEEE Radar Conference (RadarConf)*, 2017, pp. 0508–0513.

[53] "fmscan.org," https://fmscan.org/main.php?l=4.167048&b=52.031325&csave=2&qth=, accessed: 2021-02-26.

[54] ETSI. (2021) European telecommunications standards institute. [Online]. Available: https://www.etsi.org/

[55] W. Fischer, *Digital video and audio broadcasting technology: a practical engineering guide.* Springer Science & Business Media, 2008.

[56] H. Griffiths, "Bistatic and multistatic radar," in *IEE Military Radar Seminar, Shrivenham, UK*, 2004.

[57] H. Griffiths, C. Baker, H. Ghaleb, R. Ramakrishnan, and E. Willman, "Measurement and analysis of ambiguity functions of off-air signals for passive coherent location," *Electronics Letters*, vol. 39, no. 13, pp. 1005–1007, 2003.

[58] I. T. Union, "Polarization of emissions in frequency-modulation broadcasting in band 8 (VHF)," International Telecommunication Union, Tech. Rep., 1990. [Online]. Available: https://www.itu.int/dms_pub/itu-r/opb/rep/R-REP-BS.464-5-1990-PDF-E.pdf

[59] C. Bongioanni, F. Colone, T. Martelli, R. D'angeli, and P. Lombardo, "Exploiting polarimetric diversity to mitigate the effect of interferences in FM-based passive radar," in *11-th INTERNATIONAL RADAR SYMPOSIUM.* IEEE, 2010, pp. 1–4.

[60] A. J. Kamis, E. Walton, and F. Garber, "Radar Target Identification Techniques Applied to a Polarization Diverse Aircraft Data Base." OHIO STATE UNIV COLUMBUS ELECTROSCIENCE LAB, Tech. Rep., 1987.

[61] M. Borgerding, "Turning overlap-save into a multiband mixing, downsampling filter bank," *IEEE Signal Processing Magazine*, vol. 23, no. 2, pp. 158–161, 2006.

[62] L. Pucker, "Channelization techniques for software defined radio," in *Proceedings of SDR forum conference*, 2003, pp. 1–6.

[63] F. Harris, "The discrete fourier transform applied to time domain signal processing," *IEEE Communications Magazine*, vol. 20, no. 3, pp. 13–22, 1982.

[64] P. Krysik, Z. Gajo, J. S. Kulpa, and M. Malanowski, "Reference channel equalization in FM passive radar using the constant magnitude algorithm," in *2014 15th International Radar Symposium (IRS)*, 2014, pp. 1–4.

[65] H. Kuschel, M. Ummenhofer, D. O'Hagan, and J. Heckenbach, "On the resolution performance of passive radar using DVB-T illuminations," in *11-th INTERNATIONAL RADAR SYMPOSIUM*, 2010, pp. 1–4.

[66] O. Mahfoudia, F. Horlin, and X. Neyt, "Optimum reference signal reconstruction for DVB-T based passive radars," in *2017 IEEE Radar Conference (RadarConf)*, 2017, pp. 1327–1331.

[67] J. Treichler and B. Agee, "A new approach to multipath correction of constant modulus signals," *IEEE Transactions on Acoustics, Speech, and Signal Processing*, vol. 31, no. 2, pp. 459–472, 1983.

[68] J. Benesty and P. Duhamel, “Fast constant modulus adaptive algorithm,” *IEE Proceedings F - Radar and Signal Processing*, vol. 138, no. 4, pp. 379–387, Aug 1991.

[69] R. Cardinali, F. Colone, C. Ferretti, and P. Lombardo, “Comparison of clutter and multipath cancellation techniques for passive radar,” in *2007 IEEE Radar Conference*. IEEE, 2007, pp. 469–474.

[70] K. Sahu and R. Sinha, “NORMALIZED LEAST MEAN SQUARE (NLMS) ADAPTIVE FILTER FOR NOISE CANCELLATION,” *International Journal of Proresses in Engineering, Management, Science and Humanities*, vol. 1, 2015.

[71] N. Bershad, “Analysis of the normalized LMS algorithm with Gaussian inputs,” *IEEE Transactions on Acoustics, Speech, and Signal Processing*, vol. 34, no. 4, pp. 793–806, 1986.

[72] M. E. DomĂnguez, W. Hernandez, and G. Sansigre, “General block lms algorithm,” in *2009 35th Annual Conference of IEEE Industrial Electronics*, 2009, pp. 3371–3374.

[73] F. Pignol, F. Colone, and T. Martelli, “Lagrange-Polynomial-Interpolation-Based Keystone Transform for a Passive Radar,” *IEEE Transactions on Aerospace and Electronic Systems*, vol. 54, no. 3, pp. 1151–1167, 2018.

[74] Z. Gao, R. Tao, Y. Ma, and T. Shao, “DVB-T Signal Cross-Ambiguity Functions Improvement for Passive Radar,” in *2006 CIE International Conference on Radar*, 2006, pp. 1–4.

[75] A. Ludloff, *CFAR-Methoden*. Wiesbaden: Vieweg+Teubner Verlag, 2002, pp. 341–373. [Online]. Available: https://doi.org/10.1007/978-3-322-99555-1_9

[76] J. Gamba, *Radar Signal Processing for Autonomous Driving*. Springer Nature Singapore, 2020.

[77] T. Schanze, “Sinc interpolation of discrete periodic signals,” *IEEE Transactions on Signal Processing*, vol. 43, no. 6, pp. 1502–1503, 1995.

[78] M. Krueckemeier, F. Schwartau, S. Paul, and J. Schoebel, “Passive Radar Transmitter Localization Using a Planar Approximation,” *IEEE Transactions on Aerospace and Electronic Systems*, pp. 1–1, 2021.

[79] M. Malanowski, K. Kulpa, M. Żywek, and M. Wielgo, “Estimation of Transmitter Position Based on Known Target Trajectory in Passive Radar,” in *2020 IEEE International Radar Conference (RADAR)*. IEEE, 2020, pp. 506–511.

[80] M. Daun, U. Nickel, and W. Koch, “Tracking in multistatic passive radar systems using DAB/DVB-T illumination,” *Signal Processing*, vol. 92, no. 6, pp. 1365–1386, 2012.

[81] M. Edrich, A. Schroeder, and F. Meyer, “Design and performance evaluation of a mature FM/DAB/DVB-T multi-illuminator passive radar system,” *IET Radar, Sonar & Navigation*, vol. 8, no. 2, pp. 114–122, 2014.

[82] H. Kuschel, J. Heckenbach, D. O’Hagan, and M. Ummenhofer, “A hybrid multi-frequency passive radar concept for medium range air surveillance,” in *2011 Microwaves, Radar and Remote Sensing Symposium*. IEEE, 2011, pp. 275–279.

[83] J. Yi, X. Wan, Y. Fu, and G. Fang, “ADS-B information based transmitter localization in passive radar,” in *2014 XXXIth URSI General Assembly and Scientific Symposium (URSI GASS)*. IEEE, 2014, pp. 1–4.

[84] B. Kovell, B. Mellish, T. Newman, and O. Kajopaiye, "Comparative analysis of ADS-B verification techniques," *The University of Colorado, Boulder*, vol. 4, 2012.

[85] J. Nocedal and S. Wright, *Numerical optimization.* Springer Science & Business Media, 2006.

[86] M. Malanowski and K. Kulpa, "Two Methods for Target Localization in Multistatic Passive Radar," *IEEE Transactions on Aerospace and Electronic Systems*, vol. 48, no. 1, pp. 572–580, 2012.

[87] J. Smith and J. Abel, "The spherical interpolation method of source localization," *IEEE Journal of Oceanic Engineering*, vol. 12, no. 1, pp. 246–252, 1987.

[88] J. Smith and J. Abel, "Closed-form least-squares source location estimation from range-difference measurements," *IEEE Transactions on Acoustics, Speech, and Signal Processing*, vol. 35, no. 12, pp. 1661–1669, 1987.

[89] "Den Haag/Alticom Toren," https://www.senderfotos.de/senderfotos_international/niederlande/den-haagalticom-toren/, accessed: 2020-06-10.

[90] S. Saillant, P. Dorey, and S. Azarian, "Using VHF Navigation Aid to Estimate Radar Cross Section of Airliners by Inverting the Radar Equation," in *2018 International Conference on Radar (RADAR)*, 2018, pp. 1–6.

[91] R. Geise, R. Piesiewicz, A. Enders, and A. Schwithal, "Feasibility study on scaled bistatic RCS measurements of aircraft in W-band to investigate misguidance by the Instrument-Landing-System," in *2008 IEEE Antennas and Propagation Society International Symposium*, 2008, pp. 1–4.

[92] "OpenStreetMap," https://www.openstreetmap.org, accessed: 2021-04-26.

[93] M. Schäfer, M. Strohmeier, V. Lenders, I. Martinovic, and M. Wilhelm, "Bringing up OpenSky: A large-scale ADS-B sensor network for research," in *IPSN-14 Proceedings of the 13th International Symposium on Information Processing in Sensor Networks.* IEEE, 2014, pp. 83–94.

[94] "Wikipedia Liste der Offshore Windparks," https://de.wikipedia.org/wiki/Liste_der_Offshore-Windparks, accessed: 2021-02-26.

[95] "Airborne CollisionAvoidance System(ACAS) Manual," https://www.icao.int/Meetings/anconf12/Document%20Archive/9863_cons_en.pdf, accessed: 2021-04-26.

[96] I. Frede, "Evaluierung / Implementierung eines Kalman-Filters zum Zieltracking für ein 14 GHz Kollisionswarnradar," Diploma thesis, TU Braunschweig Institut für Hochfrequenztechnik, 09 2020.

[97] M. Wenzel, "Entwicklung und Charakterisierung einer elektronischen Strahlschwenkung für ein 14 GHz FMCW Radar," Master's thesis, TU Braunschweig Institut für Hochfrequenztechnik, 12 2019.

[98] F. Schwartau, "Analyse und Aufbau eines frequenzmodulierten Dauerstrichradars," Master's thesis, TU Braunschweig Institut für Hochfrequenztechnik, 10 2014.

[99] T. Instruments, "Phase Synchronization of Multiple PLL Synthesizers Reference Design," Texas Instruments, Tech. Rep., 2017. [Online]. Available: http://www.ti.com/lit/ug/tidud11/tidud11.pdf

[100] R. E. Best, *Phase Locked Loops - Design, Simulation, and Applications.* McGraw Hill Professional, 2007.

[101] P. Su and S. Pamarti, “Fractional-$N$ Phase-Locked-Loop-Based Frequency Synthesis: A Tutorial,” *IEEE Transactions on Circuits and Systems II: Express Briefs*, vol. 56, no. 12, pp. 881–885, Dec 2009.

[102] “Abwehr von unbemannten Flugobjekten für BOS (AMBOS),” https://www.fkie.fraunhofer.de/de/forschungsabteilungen/kom/ambos.html, accessed: 2021-04-20.

# Publications

- **Markus Krueckemeier**, Fabian Schwartau, Sebastian Paul and Joerg Schoebel
  "*Passive Radar Transmitter Localization Using a Planar Approximation*"
  IEEE Transactions on Aerospace and Electronic Systems, vol. 57, no. 5, pp. 3405-3415, Oct. 2021
- Sebastian Paul, Fabian Schwartau, **Markus Krueckemeier**, Reinhard Caspary, Carsten Monka-Ewe, Joerg Schoebel and Wolfgang Kowalsky
  "*A Systematic Comparison of Near-Field Beamforming and Fourier-Based Backward-Wave Holographic Imaging*"
  IEEE Open Journal of Antennas and Propagation, vol. 2, pp. 921-931, 2021
- **Markus Krueckemeier**, Fabian Schwartau and Joerg Schoebel
  "*A modular localization system combining passive RF detection and passive radar*"
  Kleinheubacher Tagung 2019 (KH 2019), Miltenberg, Germany
- **Markus Krueckemeier**, Fabian Schwartau, Carsten Monka-Ewe and Joerg Schoebel
  "*Synchronization of Multiple USRP SDRs for Coherent Receiver Applications*"
  6th International Conference on Software Defined Systems (SDS 2019), Rome, Italy
- Fabian Schwartau, Stefan Preussler, **Markus Krueckemeier**, Florian Pfeiffer, Hannes Stuelzebach, Thomas Schneider and Joerg Schoebel
  "*Modular Wideband High Angular Resolution 79 GHz Radar System*"
  12th German Microwave Conference (GeMiC 2019), Stuttgart, Germany
- Fabian Schwartau, Carsten Monka-Ewe, **Markus Krueckemeier** and Joerg Schoebel
  "*Aircraft window attenuation measurements at 60 GHz for wireless in-cabin communication*"
  11th German Microwave Conference (GeMiC 2018), Freiburg, Germany

# Abstract

Airspace surveillance is one of the classic applications of radar technology. Existing systems are aimed at large-scale surveillance of commercial airspace. In this thesis, a prototype of a smaller, mobile, passive radar system is presented, which can be used, for example, to monitor aircraft and UAV in the airspace over maritime borders. The focus is on the application of the system as a passive radar system. The big advantage of passive radar systems is that they do not need their own illuminator signals and instead use existing radio systems. The system's hardware is designed as a passive coherent receiver system. Such a system can also be used to detect UAVs directly via their telemetry signals.

This thesis shows first the development of a versatile and mobile coherent receiver system that includes all components from the antenna arrays to the computer for signal processing. The presented system is designed as an experimental platform for passive detection experiments and, if required, can be modularly extended with further receiver arrays with various geometries and frequencies.

On the basis of this system, a complete signal processing is presented, from the recording of digitalized signals to tracked positions and velocities of targets. To calculate the target track from the recorded measurements, information about the transmitter's positions is needed in addition. In order not to be dependent on external databases, a method is proposed that makes it possible to determine this information from the measurement results of the system in combination with the ground truth information that is recorded via ADS-B from cooperative targets. Additionally, a model of the detection performance is developed that allows predictions to be made about the coverage. The function of the system, the developed signal processing, and the performance model are proven with measurements of a real scenario in the Netherlands.

# Kurzfassung

Die Luftraumüberwachung ist eine der klassischen Anwendungen der Radartechnik. Bestehende Systeme sind auf die großflächige Überwachung des kommerziellen Luftraums ausgerichtet. In dieser Arbeit wird ein Prototyp eines kleinen, mobilen, passiven Radarsystems vorgestellt, das z.B. zur Überwachung von Flugzeugen und UAVs im Luftraum über Seegrenzen eingesetzt werden kann. Der Schwerpunkt liegt dabei auf der Anwendung des Systems als Passivradarsystem. Der große Vorteil von passiven Radarsystemen ist, dass sie keine eigenes Sendesignal zur Beleuchtung der Ziele benötigen und stattdessen vorhandene Funksysteme nutzen. Die Hardware des Systems ist als passives kohärentes Empfangssystem ausgelegt. Ein solches System kann auch verwendet werden, um UAVs direkt über ihre Telemetriesignale zu erkennen.

Diese Arbeit zeigt zunächst die Entwicklung eines vielseitigen und mobilen kohärenten Empfängersystems, das alle Komponenten von den Antennenarrays bis zum Computer für die Signalverarbeitung umfasst. Das vorgestellte System ist als experimentelle Plattform für passive Detektionsexperimente konzipiert und kann bei Bedarf modular um weitere Empfängerarrays mit unterschiedlichen Geometrien und Frequenzen erweitert werden.

Auf der Basis dieses Systems wird eine komplette Signalverarbeitung vorgestellt, von der Aufzeichnung der digitalisierten Signale bis hin zur Verfolgung der Position und Geschwindigkeit der Ziele. Um die Trackinformationen der Ziele aus den aufgezeichneten Messungen zu berechnen, werden zusätzlich Informationen über die Position der Sender benötigt. Um nicht von externen Datenbanken abhängig zu sein, wird eine Methode entwickelt, die es ermöglicht, diese Informationen aus den Messergebnissen des Systems in Kombination mit den Ground-Truth-Informationen, die über ADS-B von kooperativen Zielen aufgezeichnet werden, zu bestimmen. Zusätzlich wird ein Modell der Detektionsperformanz entwickelt, das Vorhersagen über die Abdeckung ermöglicht. Die Funktion des Systems, die entwickelte Signalverarbeitung und das Leistungsmodell werden mit Messungen eines realen Szenarios in den Niederlanden nachgewiesen.

www.ingramcontent.com/pod-product-compliance
Ingram Content Group UK Ltd.
Pitfield, Milton Keynes, MK11 3LW, UK
UKHW021653190726
13853UKWH00001B/229

9 783736 976740